Derivatives Algorithms

Volume 1: Bones   Second Edition

Derivatives Algorithms

Volume 1: Bones   Second Edition

Tom Hyer

World Scientific

NEW JERSEY · LONDON · SINGAPORE · BEIJING · SHANGHAI · HONG KONG · TAIPEI · CHENNAI · TOKYO

Published by

World Scientific Publishing Co. Pte. Ltd.

5 Toh Tuck Link, Singapore 596224

USA office: 27 Warren Street, Suite 401-402, Hackensack, NJ 07601

UK office: 57 Shelton Street, Covent Garden, London WC2H 9HE

Library of Congress Cataloging-in-Publication Data
Hyer, Tom.
   Derivatives algorithm : Volume 1: bones / Tom Hyer, UBS, UK. -- Second Edition.
     pages cm
   Includes bibliographical references and index.
   ISBN 978-9814699518 (hardcover : alk. paper)
   1. Derivative securities--Data processing. 2. Algorithms--Data processing. I. Title.
   HG6024.A3H94 2015
   332.64'57028551--dc23

                                 2015024082

British Library Cataloguing-in-Publication Data
A catalogue record for this book is available from the British Library.

In-house Editors: Rajni Gamage/Karimah

Typeset by Stallion Press
Email: enquiries@stallionpress.com

Printed in Singapore

For Kath – first, last, and always

vi   *This page intentionally left blank*

Contents

Chapter 1

Introduction

In June 1994, I spent my last thousand dollars on two suits and a plane ticket to New York. I landed a job in Fixed Income Analytics at Bankers Trust; I had hoped for an offer from Enron, but that was slow to materialize. I quickly realized that I had been hired to produce not mathematics but algorithms.

After Bankers Trust was absorbed into Deutsche Bank, I left for First Union and then for UBS. As I moved into positions of broader responsibility in more successful businesses, the primacy of algorithms became still more pronounced, and I adopted the title of this book as a kind of slogan and began using it on my business cards.

Yet the books available to current and potential practitioners are almost exclusively about mathematics. Many of these books are excellent, but by their nature they can say little about the reality of most quant life. I will try to write about our real work – about the issues that must fill the attention of a library-building quant.

The result will inevitably be shaped by my own preconceptions and areas of ignorance. I can give any deconstructionists a head start by disclosing my perspective. I have been the lead modeller and coder for groups whose main function is to provide a library of analytic functions (models, calibrations thereof, and trade pricing engines) for derivatives pricing. The "desk quants" who study individual trades, and the traders who take final responsibility for the pricing, are my clients. My group's job is to extend the range of what is possible – to give a structurer in Hong Kong the power to price a complex hybrid trade, using tools he understands well enough, roughly the way he might understand his car. I am also accustomed to working within a group, and creating protocols that are effectively binding on other coders. The hobbyist pricing complex derivatives in his spare time may find the resulting code too rigid.

1

Finally, much is omitted here. I do not intend to share insights into any firm's current operations, nor to display the models I consider best. Nor will I betray the reader by publishing castoffs, the failed experiments along the road to working models. Where I cannot write honestly, I will remain silent. Thus the current volume, as its subtitle suggests, demonstrates the creation only of a framework in which powerful models can be used to price very general trades.

1.1 Note on the Second Edition

This second edition follows the path of its predecessor, but incorporates new material and several changes in presentation.

Chapters 5 and 10 have been substantially rewritten, with more realistic code and, in the former case, the support of a publicly available code generator. Chapter 12 has an updated treatment of yield curves and funding which is far better suited to today's markets. And a new Chapter 16, describing the next phase in the evolution of derivatives valuation and risk, has been added.

The code examples have been substantially updated. They use new features introduced with the C++11 standard, and some functions and classes have been renamed to improve clarity. Some special-purpose classes have been replaced with widely available alternatives (*e.g.*, we now use `boost::optional` rather than `Maybe_`). Finally, some implementation errors have been corrected, though the example code is still meant to be didactic rather than functional.

Where mark-up for code generation is used, it is now formatted in the style expected by Machinist (see Sec. 3.5), a public-domain code generator.

The contents have been rearranged in several ways:

- Exception messaging no longer uses the `ENV` mechanism.
- The archiving code has been adapted to be compatible with Machinist, and `Dst_`/`Src_` have been renamed to the less crpytic `Store_` and `View_`, respectively. On the whole the persistence mechanism is less powerful than that described in the first edition, but simpler to implement and extend, and we present it in more detail.
- The appendices containing extended code samples have been removed; instead, the code is available in a Bitucket repository (`http://bitbucket.org/hyer/DA_Bones`); contact the author at `DerivativesAlgorithms@gmail.com` for details.

It is important to understand that the online repository of code is meant for explanation rather than for production use, and is designed to support this book. Some concepts are only sketched, rather than brought to completion; parts of a working library which are not directly addressed in this volume, such as the generation of fixing and payment schedules for swaps and bonds, may well remain unfinished indefinitely. The online documentation at Bitbucket is sparse as well; explaining the code is, after all, the point of this book.

The book's blog (http://derivativesalgorithms.blogspot.com/) will be updated with errata and extensions.

Finally, as my career is no longer centered on derivatives pricing, I am no longer fully immersed in the concepts this work describes. Thus I have been able to add some perspective and explanation, making this edition more readable and accessible.

This page intentionally left blank

Chapter 2

Principles

This is not a book about mathematical methods, nor does it explain any class of models in real detail. (For that purpose, consider the *magnum opus* of Leif Andersen and Vladimir Piterbarg, on *Interest Rate Modeling*.) Rather, it is an exposition of the methods needed to describe derivatives and structured financial products in a precise and flexible way, so that both their innate complexity and the complexity of the models for pricing and hedging them them can be controlled. Separation of concerns is a crucial part of this control, and one of our major aims is to demonstrate where this separation may best be accomplished. We also concentrate on the development of reusable components, so we can always understand a given program in terms of large and well-understood atoms; and on a style which encourages clear and concise expression of our intentions.

2.1 Our Code

Most of the code examples presented here will be given in C++. This is partly a result of the author's own experience. In addition, it reflects the expectation that most of the code described will form part of a function library, for potential use by many different applications, and C++ is the library-design language *par excellence*. We will generate a substantial part of this C++ code from higher-level inputs (see "Interface Generation", below); the same techniques will be used to construct interfaces to other languages. We will often describe some concept by showing this *mark-up code* instead of the generated C++ code. The C++ standard library will be used freely, and without namespace qualifiers; however, its different stylistic conventions make these contributions easy to spot.

The code examples will inevitably embed our own conventions. The most important of these is that we avoid passing non-const references, because they obscure the meaning of user code: does f(x) change the value of x? This differs sharply from the C++ standard library convention; in particular, we will write a1.Swap(&a2).[1] We also attempt to add information through formatting, using case and underscores to distinguish the names of Classes_, Functions, memberData_, localVariables and function_arguments.

When a function is nonstandard in some important way or is explicitly *not* part of an interface, we will put an X in front of its name. Think of X as standing for "exclusively expert," and signifying that the function should not be called without understanding its implementation. The most common use of this notation is for *ephemeral classes* which hold references whose duration cannot be guaranteed.

The layout of example code is constrained by the 66-column width of the printed page, less than half that of an ordinary modern development screen. I have sometimes compensated for this by using artificially short variable names, which I hope will still convey enough of their purpose to the reader. I have also introduced the macro

```
#define DYN_PTR(n, t, s) t* n = dynamic_cast<t*>(s)
```

solely to reduce line lengths.

2.1.1 auto

Typenames in C++ can be quite long, and sometimes depend on implementation details of template libraries. The auto keyword, introduced in the C++11 standard, saves effort in several ways. The code examples in this book use auto extensively.

2.1.2 override *and* final

In C++, marking a member function as virtual creates an entry in the enclosing class's *vtable*; unless the function is itself an override of a base class's virtual function, in which case it has no effect. So a sequence of virtual functions has a distinctive beginning (the virtual keyword in the base class) but no other markers.

---

[1] We will break our own rule at times, when using signaling classes whose whole purpose is to be altered.

The C++11 standard introduces keywords to indicate intention for `virtual` functions; `override`, which indicates that a function implements a base class's `virtual` function, and `final`, which forbids override in a further derived class. (Classes can also be marked `final` in their totality.) We use the `override` keyword wherever appropriate,[2] since it lets the compiler detect signature mismatches (*e.g.*, omitting `const` from an overriding function). We are less certain of the value of `final`, since we may wish to add instrumentation even to fully-implemented classes, and we do not use it in our examples here.

2.1.3 *Wishful Thinking*

We will declare and define constructors using a syntax which is not part of C++, though we believe it should be. To wit: when we use a member datum name in place of a parameter to the constructor, we mean that the argument should be of a `const` reference to the member's type (or potentially, of any type from which the member can be constructed) and that it shall be used to construct that argument. For example,

```
   class Circle_
   {
       double radius_;
       Point_ center_;
 5     // we write this:
       Circle_(radius_, center_) {}
       // we mean this:
       Circle_(const double& radius, const Point_& center)
       :
10     radius_(radius),
       center_(center)
       {   }
       // we write this:
       Circle_(radius_, center_...) {}
15     // we mean this:
       template<typename... Args_> Circle_
          (const double& radius, Args_ &&... center)
       :
       radius_(radius),
20     center_(std::forward<Args_>(center)...)
       {   }
   };
```

---

[2]We will occasionally omit `override` in examples, simply to fit this format's line length.

This keeps our code examples more compact and readable, and can easily be translated back into legal C++ by the reader.

2.2 Functional Programming

Our programs will be written in a largely functional style, with most variables and objects being immutable after their construction. This approach is the fundamental strength of "pure" functional languages like Haskell; we make no attempt at purity, but a modicum of discipline provides a great increase in code reusability and maintainability.

The first gain from this style is *referential transparency*, meaning that at each level of the code we need only understand what a function *returns*, not what it *does*. Since errors often arise from obscure conditions encountered during large batch runs in distant lands, this practice simplifies many urgent debugging tasks. This style also leads naturally to code which is less brittle, so that an implementation change is less likely to have unforeseen ill effects.

Within a function body, we will attempt to use `const` values, and to declare them as such. However, we have little compunction about using iterators and such imperative constructs inside functions when they are convenient.

2.3 Type and State

An unfortunately common coding practice is to create an object during one phase of processing, then later to "activate" it in preparation for some anticipated use. For example, a parsed expression from a trade-description language might be activated to prepare for repeated evaluation in different model-generated scenarios. This is invariably a bad practice.

It is easy to see the minor problem: the object before activation is complicated by the empty slots which will be populated during activation, while the object after activation is still carrying around the original data. But this is merely a symptom of a deeper error, which is that the two phases of the object's existence are really two different things, wrongly jammed together in a single object. Another symptom is the absence at compile time of information about the object's state, creating the risk of failure-to-activate.

The solution is now obvious: activation is not a transformation of the initial object, but the creation of a new object. This fits within our functional programming paradigm, and brings the concomitant benefits: referential transparency, more freedom of optimization, compiler assistance in detecting many errors, and more fine-grained separation of concerns. We will see this pattern repeatedly.

2.4   Physical Code Structure

Regardless of language, separation of concerns requires support from the physical structure of the underlying code. If one piece of code "knows about" another – *e.g.*, in C++, if it includes the other's header, however indirectly – then it probably cannot be understood, and certainly cannot be linked, tested or debugged, without the latter. Additionally, each user of the former code perforce "knows about" the latter in the same sense, making changes less local and greatly extending the compilation and link cycles.

We will always seek to minimize such dependencies. It is common, especially in object-oriented programming, to conceal a class's implementation from its clients; however, we should go further and to the greatest possible extent should conceal the *existence* of classes that are not directly needed. This gives rise to a pattern of separation, at each level of the code: at the "upper" level are objects that unite several members to perform a coherent task, using lower-level functions that are unaware of the object's existence. It also leads to an emphasis on protocols: upper-level objects use abstract classes embodying lower-level protocols, without any dependency on the concrete classes which implement those protocols. Since we are writing a library, not an application, we will never think in terms of a "top" level. Any attempt to create a top level – *e.g.*, the creation of a static list of possible implementations of a protocol – should be viewed as a bug.[3]

An important part of minimizing dependencies is to keep information out of header files when possible. We will rely on *local classes*, implemented within a local namespace in a single source file, with only a factory function to create a *new* instance visible in the header. As a rule of thumb, the aggregate size of source files should be at least six times that of headers in a passably maintained code base.

---

[3]Our aversion to `final` is a milder statement of this same principle.

2.4.1   *Facts*

Our mindset is toward the development of a functional library, but realistically we must expect to export some information to the user as well. Examples include:

- Lists of known holidays – see Sec. 8.2
- Schedule defaults per currency – see Sec. 8.3
- Bonds deliverable against a bond future

These are part of the computational environment, but it does not really make sense to speak of them as user environment; they are not subject to the user's choice, but are facts about the external world we are modeling.

Such data can be loaded into the library in a plethora of ways, from hardwiring in the code to run-time registry of data sources. The only unifying principle is that we should separate the data acquisition from its use. Thus we do not form a dependency of the data user on the data source, which would reward hardcoding and restrict implementation freedom. Instead, both the data user and an independent data supplier depend on a low-level data storage component.[4]

2.5   Platform

In C++, the contents of a given source file should be described[5] in a header file of the same name, and that header should be *self-contained*: including it should not require the inclusion of some other header. For example, suppose a class `Slide_` is defined in `Slide.h`, and then `SlideEquity.h` defines a class derived from `Slide_`, but that this new header file cannot be compiled without first including `Slide.h`. If we write in `SlideEquity.cpp`

```
#include "Slide.h"
#include "SlideEquity.h"
```

then it will compile without complaint, but a higher-level file which includes `SlideEquity.h` may be confronted with an irritating and unexpected error (which can only be corrected by including `Slide.h` in that file as well).

The usual way to police this is to have each source file include its own header before including other headers, which might hide dependencies. We

---

[4] Sec. 8.3 shows an example data supplier.

[5] To the extent that they must be available outside that source file.

adopt the policy that each source file must include its own header file as the *second* line of code: the first line of code is reserved for the *platform header* Platform.h. The platform header thus provides the minimal shared development environment for all code. As such, it has several responsibilities:

(1) Forward declare common STL/Boost/Loki classes and interface classes.
(2) Apply using declarations and directives to be used everywhere.
(3) Supply some very simple macros, *e.g.* UNREACHABLE and EXPORT.
(4) Optionally, supply basic math functionality, *e.g.* Square and IsZero.

Some of the items listed have their own complications. A surprising example is that forward declaration of standard library templates is explicitly forbidden by the C++ standard, even though no good substitute is offered (except the one case of iosfwd.h). We do it anyway, on platforms where it can be made to work.

Our preferred definition of UNREACHABLE,

Platform.h
```
#define UNREACHABLE assert(!"Reached unreachable code")
```

depends on a system-defined assert which must be included. It is a good practice to separate host-specific code into its own header files, so Platform.h might contain:

Platform.h
```
   // ...
   #ifdef _MSC_VER
   #include "DA_MSWindows.h"
   #endif
 5 #ifdef __linux__
   #include "DA_Linux.h"
   #endif

   // host-independent code
10 class Dictionary_;
   template<class T_> class Vector_;
   // ...
```

Our use of "host" to describe the underlying compiler and hardware, rather than the usual "platform", is deliberate: we wish to emphasize that Platform.h creates the unique platform on which we will develop.

We also use a second system-wide header, Strict.h, to enforce coding standards which are not necessarily shared by third-party libraries. One of its roles is to promote warnings to errors: some C++ warnings are in fact almost always indications of true errors, and we will not accept them anywhere in our code. With Microsoft's compiler, at a minimum we would promote

- 4150, deletion of pointer to incomplete type;
- 4800, forcing value to bool 'true' or 'false' (this often means an unexpected overloaded function or operator is being invoked);
- 4172, returning address of local variable or temporary;
- 4717, function recursive on all control paths;
- 4129, unrecognized escape sequence;
- 4244, conversion from float to integer type.

In exchange for the suppression of float-to-int conversion, `Strict.h` also provides a function `AsInt` which supplies a warning-free truncation to integer. We also disable some otiose warnings; *e.g.*, for Microsoft, 4786 (symbol greater than 255 characters).

Our practice is to `#include Strict.h` in each source file, after any system headers, but before other headers in our own library.

2.6   Some Design Patterns

A few "design patterns" (in the sense of Gamma *et al.*) do occur repeatedly in our work. These are general programming techniques which, with a little adaptation, can often provide good solutions to a wide variety of design problems. Their use in derivatives pricing has been discussed by Mark Joshi in some detail,[6] and his book is highly recommended to the interested reader.

2.6.1   *Factory Method*

A factory method is a free function which creates a concrete instance of an abstract class, without disclosing the definition of the concrete class. We will use this to conceal (within a source file) the definition of almost every complicated concrete class.

2.6.2   *Decorator*

A decorator is a class, derived from some abstract interface class, which holds another instance derived from that same interface. It can implement

---

[6]*C++ Design Patterns and Derivatives Pricing*, Cambridge University Press, 2004.

each part of the interface itself, or by forwarding to the member instance, or by forwarding and then "tweaking" the result. The clearest example is a bumped discount curve, which stores the base curve as a member. It computes discount factors by obtaining them from the base curve, then making any necessary adjustment for the bump in rates.

A class cannot safely be decorated if it has *impure* virtual functions; there is too much risk that the decorator will fail to forward the call (*e.g.*, if the base class function's signature changes). For such classes, or if decoration is a common task, we will construct a *null decorator* which simply forwards everything to its internal instance. Each decorator then inherits from the null decorator, not directly from the base class; see Sec. 11.9.1 for an example.

2.6.3 *Singleton*

A singleton is an object which is guaranteed to exist when needed and to be unique within a running application or process. Run-time registries, which are invaluable in reducing compile-time dependencies, must be singletons. We almost always use the Meyers singleton (a static object within an accessor function which returns a reference thereto). As a rule, registries are populated at load time. We provide the macro `RETURN_STATIC(...)`[7] which declares and returns a `static` object of the given type; this is the function body of a Meyers singleton. We usually begin the singleton function's name with `The`.

We do not use non-`const` `static` objects for any purpose except singletons.

2.6.4 *Composites*

A composite is a set of instances of a class, which collectively form another instance of that same class. We implement this with a template class:

```
――――――――――――――― Composite.h ―――――――――――――――
template<class T_, class H_ = ElementHolder_<T_>::type>
class Composite_ : public remove_const<T_>::type
{
public:
    typedef H_ element_t;
protected:
    Vector_<element_t> contents_;
```

―――――――――――――――――――――――――――――

[7]A variadic macro, in case the type argument contains commas.

The use of `remove_const` allows us to specify a `const` type for `T_`, so the components within the composite will be individually `const`. The `ElementHolder_` is a traits class, defaulting to `shared_ptr` for non-`const` and `Handle_` for `const` elements.

It may be that the base type `T_` itself lacks a default constructor. Using C++11's variadic templates, we can still implement `Composite_<T_>` and provide appropriate constructors:

```
_____ Composite.h _____
    template<typename... Args_> Composite_(Args_&&... args)
        : T_(std::forward<Args_>(args)...) {}


    template<typename... Args_> Composite_
5       (const Vector_<element_t>& contents, Args_&&... args)
        : T_(std::forward<Args_>(args)...), contents_(contents) {}
```

These two functions construct an empty composite, and a composite from input elements, respectively.

2.7 Optimization

Efficiency of code is an important consideration, which can never be put completely out of mind. However, the sacrifices we are willing to make for efficiency are limited: clarity, functional style, and transparency of intentions are all incommensurably more important in most of our code. Only in known code hotspots will we focus strongly on efficiency.

It is also worth mentioning that the dominant source of performance degradation in the field is unnecessary memory allocation, sometimes with the addition of copying. Fine-tuning inner loops is almost always useless; instead, developers should focus on pro-actively constructing and then reusing immutable objects which carry the information a computation needs.

2.7.1 *Calibration*

It turns out that in practice we pay almost no penalty for this cavalier approach, except in one area: calibration. Because it inevitably entails the creation of high-level objects for each candidate solution, our preference for immutable objects entails more allocation and object initialization than would a more imperative-minded approach. In particular cases like high-frequency trading, we may be forced away from our principles.

It is valuable to distinguish these problem areas as exceptions to be carefully quarantined, rather than imagining that they form a general rule. My own repeated experience has been that "efficiency" is used as a mantra to defend many poor programming practices.

2.7.2 map

A "map", or associative container, is a set of unique keys and an associated value for each; but a `map` is a particular C++ template class, which implements a sorted associative container using a red-black tree. For the most efficient code, we must understand the variety of available containers:

- `map` – insertion, deletion, iteration and lookup are all asymptotically acceptably fast ($O(\lg n)$); but, because each node is heap-allocated, the constant factor is large.
- `hash_map` – lookup is substantially faster than a `map`'s, but heap allocation remains a problem. Elements are not sorted, so there is no concept of `lower_bound`.
- `AssocVector` – from Alexandrescu's Loki library. Internally, this is a sorted `vector`, so heap allocation is rarer but insertion can take $O(n)$ time. Iteration is optimal; lookup is as fast as the comparison function will allow.

In this work, we use `map` everywhere by convention. This does not affect the code's correctness, but production code would make different choices for efficiency.

2.8 Threads

Much of the code we will display here is not thread-safe; for example, we use a static repository of in-process objects. In production code, we should guard any access to such an object with a *mutex* – an object supplied by any threading library, which once constructed by one thread will block other threads in its constructor, forcing them to wait until the first thread releases the mutex. In C++, mutexes are inevitably implemented using RAII ("resource acquisition is initialization") – they are locked on construction of a "lock object", and freed in its destructor.

We should obviously avoid sharing non-`const` objects between threads – a corollary is that `mutable` data members, such a cache of results or a vector of workspace, should not be used around threads. To price a path-dependent trade in a multithreaded Monte Carlo, for instance, we must maintain a separate instance of the trade's state for each thread. All our code examples are single-threaded, but should be "future-proof" when (non-`const`) `static` and `mutable` objects are not directly involved.

Chapter 3

Types and Interfaces

Some of the code we must produce is nearly independent of the task at hand: the jobs of translating inputs to our own formats, storing and fetching high-level objects, and communicating error messages are all crucial (though each single instance of doing so is less than fascinating). Our first task is to create a supporting system to allow these jobs to be done well, with minimal unnecessary work and boilerplate code.

3.1 The User Base

To create a functioning quant group capable of continuous improvement and re-engineering, we must demarcate what is and is not the work of quants. If pricing tasks are diffused across too large a group, then pricing code will be written in many different locations and will become nearly impossible to track down and change; similarly, a quant group with an over-large mandate will inevitably come to spread pricing code over its entire domain. We avoid this by separating the quant group from its users.

This is a large part of the rationale for our focus on library production: it provides a real-world mechanism for separating pricing and risk tasks (the natural preserve of quants) from database, data transmission, distributed computing and graphical interface tasks. A corollary is that internal types used in the pricing process are the preserve of quants only, and the C++ header files used to build the library are *not* the library interface to its users, which we call the *public* interface.

The concepts we must incorporate into the interface are those which exist outside the process of pricing or risk, or form its top-level inputs: *e.g.* portfolio and risk report. As we create new ways for users to specify these, other concepts will creep into our interface: *e.g.* models, yield curves and even interpolants. But most of our C++ classes will never appear in the public interface, since we often deal in abstractions internal to our own library. The public interface must generally be backward-compatible, but header files away from the public interface may be changed by the quant group without affecting any other group.

3.2 A Public Example

The concepts of the public interface are far easier to understand by example than abstractly. Thus we begin by displaying a public function to query an interpolant. The inputs are a (pre-existing) function object and a vector of x-values at which to interpolate. We describe the function using not C++, but mark-up which will be read by our code generator:

```
_____ __Interp.cpp _____
   /*IF------------------------------------------------------
   public Interp1_Get
        Interpolate a value at specified abcissas
   &inputs
5  f is handle Interp1
        The interpolant function
   x is number[]
        The x-values (abcissas)
   &outputs
10 y is number[]
        The interpolated function values at x-values
   -IF----------------------------------------------------*/
```

This is a high-level description of a function, from which C++ code can be generated. The `Interp1` type must have some meaning in C++, which is naturally described in a header file (not part of the public interface):

```
_____ Interp1.h _____
   #pragma once

   #include "Archive.h"

5  class Interp1_ : public Storable_
   {
   public:
        Interp1_(const String_& name);
        virtual double operator()(double x) const = 0;
10      virtual bool IsInBounds(double x) const
            { return true; }
   };
```

Here `Storable_` is our base class for objects visible to library users. In an environment like Excel, such an object will appear to the user as a "handle string" which will be used to look in a `static` singleton repository of `Storable_` objects (see Secs. 3.3 and 5.6).

We wish to separate interface code from library implementation code, so we put it in a different file. Under our preferred naming convention, this source file will be called `__Interp.cpp`; we use the leading `__` to identify public-interface files.

```
———————————— __Interp.cpp ————————————
#include "__Platform.h"
#include "Strict.h"

#include "Interp.h"
```

In interface code, we include `__Platform.h` (which in turn includes `Platform.h`) as the first header. This provides definitions of input types like `pOPER`, and utility functions for translating them. Our preference is to include `Strict.h` after system headers, but before any of our own headers. This allows us to increase and customize warnings *after* including external library code which is not always warning-free.

```
———————————— __Interp.cpp ————————————
   namespace
   {
      double CheckedInterp(const Interp1_& f, double x)
      {
5        REQUIRE(f.IsInBounds(x), "X (= " + to_string(x)
              + ") is outside interpolation domain");
         return f(x);
      }

10    void Interp1_Get
         (const Handle_<Interp1_>& f,
         const Vector_<>& x,
         Vector_<>* y)
      {
15       *y = Apply([&](double x_i)
              {return CheckedInterp(*f, x_i); }, x);
      }
   } // leave local namespace

20 #include "MG_Interp1_Get_public.inc"
```

This much code is handwritten. Next we invoke an *interface generator* which scans the directory and finds the `/*IF` which marks the beginning of

20 Derivatives Algorithms

a mark-up block. It creates[1] the implementation file, which is then present for inclusion into the source.

```
──────── MG_Interp1_Get_public.inc ────────
extern "C" __declspec(dllexport) OPER_* xl_Interp1_Get
   (const OPER_* xl_f, const OPER_* xl_x)
{
   Excel::InitializeSessionIfNeeded();
5  ENV_SEED_TYPE(ObjectAccess_);
   const char* argName = 0;
   try
   {
      Log::Write("Interp1_Get");
10     argName = "f (input #1)";
      const Handle_<Interp1_> f =
            Excel::ToHandle<Interp1_>(_env, xl_f);
      argName = "x (input #2)";
      const Vector_<double> x =
15           Excel::ToDoubleVector(xl_x);
      argName = 0;

      Vector_<double> y;
      Interp1_Get(f, x, &y);
20     Excel::Retval_ retval;
      retval.Load(y);
      return retval.ToXloper();
   }
   catch (exception& e)
25 {
      return Excel::Error(e.what(), argName);
   }
   catch (...)
   {
30     return Excel::Error("Unknown error", argName);
   }
}

struct XlRegister_Interp1_Get_
35 {
   XlRegister_Interp1_Get_()
   {
      Vector_<String_> argHelp;
      argHelp.push_back("The interpolant function");
40     argHelp.push_back("The x-values (abcissas)");
      Excel::Register("Bones", "xl_Interp1_Get",
         "INTERP1.GET",
         "Interpolate a value at specified abcissas",
         "QQQ", "f,x", argHelp, false);
45  }
};
static XlRegister_Interp1_Get_ The_Interp1_Get_XlRegisterer;
```

[1]Or updates, as appropriate.

Similar code would be added to support a Java, .NET, or other foreign function interface.[2] Even if calling applications were written in C++, we would still expect them to communicate with our code only through this interface – thus defending the separation of quant from non-quant code.

The LogXLCall function performs administrative tasks such as checking for library expiration or updating a session usage log; here we deduce the DLL name from the source file path.

The outlines of this code are apparent; this chapter is devoted to explaining the design choices that lead to this particular form.

3.3 Interface Generation

At the public interface of our code library, we need to produce functions for use by applications. Since these applications are not part of our code base – for example, they may be third-party products like Microsoft Excel – the interface should not be presented in terms of our own (C++) types, or of reflections of our types into another language. We will instead have some basic types, and a generic "object" type for internal high-level concepts. Different applications will require different internal representations of the object, such as the handle strings seen by an Excel user.

Like any foreign interface, this requires interface code in our own library combined with additional *mark-up* information for each function. The latter information is not specific to a particular interface; rather, it is a high-level description in our own terms, to be read by our own interface generator.

A further advantage of interface generators is that they can generate multiple outputs. At a minimum, they can produce documentation (*e.g.*, an HTML help page) at the same time as the interface – thus ensuring that the documentation will be always up-to-date, which otherwise requires harsh control over programmers. Also, though only an Excel interface is displayed here, the same information suffices to generate interfaces to Java, .NET, or any other external platform.

The interface generator can be written in many ways; Sec. 3.5 describes a public-domain generator which is used in this work.

---

[2]If we wish to support very many such interfaces, we might interface a single protocol such as SOAP.

3.4   Interface Types

The fundamental types required should be chosen to be helpful to the user without being too onerous to support. For `public` functions, our generator supports the following types:

- `number` – a real scalar, presumably double-precision
- `integer`
- `string` – a single text string
- `boolean` – a true/false value
- `cell` – a discriminated union of number, text or boolean, like a spreadsheet cell
- `date` – a date without a time-of-day
- `datetime` – a date and time
- `dictionary` – essentially a map<String_, Cell_>
- `handle` – a Handle_ to an object of ours, opaque to the user
- `enum` – an object of ours, constructible from a string (*e.g.*, call-put flags)
- `settings` – an object of ours, constructible from a dictionary

The last three entries are not types, but families of types; thus the `handle Interp1` notation in the markup for `Interp1_Get` specifies an object of a particular type. In our code, an `enum` will map to an instance of a class with an enumerable set of valid values; *e.g.*, a coupon basis. If such classes have canonical conversions to and from strings, then they can be constructed with mechanically generated code and passed to our functions; this is detailed in Sec. 8.1.

One- and two-dimensional arrays of scalar types are also supported: thus the `number[]` type is a vector of real numbers. Arrays of `enum`, or two-dimensional arrays of `date` or `handle`, seem not to be very useful in practice.

3.4.1   *Tables and Cells*

A table is a two-dimensional matrix of cells. We will use the same template `Matrix_` class for all element types, so we focus here on defining the cell.

We no longer use Boost's `variant` template for cells, judging that the task is not complex enough to merit the compilation time required. Our

Cell_ class is a handwritten discriminated union, which can be constructed or assigned from a number, boolean, date or string; queried for its nature; and coerced to any of those types.

We can also create a general type checker (in namespace Cell):

```
────────────────────── TableUtils.h ──────────────────────
struct TypeCheck_
{
    std::bitset<static_cast<int>(Type_::N_TYPES)> ret_;

5   TypeCheck_ Add(const Type_& bit) const
    {
        TypeCheck_ ret(*this);
        ret.ret_.set(static_cast<int>(bit));
        return ret;
10  }
    TypeCheck_ String() const
        { return Add(Type_::STRING); }
    TypeCheck_ Number() const
        { return Add(Type_::NUMBER); }
15  TypeCheck_ Date() const
        { return Add(Type_::DATE); }
    TypeCheck_ DateTime() const
        { return Add(Type_::DATETIME); }
    TypeCheck_ Boolean() const
20      { return Add(Type_::BOOLEAN); }
    TypeCheck_ Empty() const
        { return Add(Type_::EMPTY); }

    bool operator()(const Cell_& c) const
25  {
        return ret_[static_cast<int>(c.type_)];
    }
};
```

This wraps the process of interconverting between type_ and its integer representation, and lets us write checking code like:[3]

```
Require(Check(correlations.Col(1),
            Cell::TypeCheck_().String().Number()),
        "Correlations must be numbers or handle strings");
```

3.4.2 *Variety*

Why so many types? We could do away with everything except cell[][] and construct everything else from it: *e.g.*, an object is extracted from the

---

[3]But without the artificially short lines.

repository using the string value of the sole cell of a 1×1 array. But by creating more types, we allow more checking to be done in the machine-generated interface layer, and thus ensure that the checking code will never be forgotten or neglected, or created by careless copy-and-paste.

Also, when exporting our API to platforms other than Excel, we profit from interaction with the user's type-checking system. A `handle` returned to a Python caller need never be placed in a static repository; instead, it can be shared with a Python object returned to the user.

In `MG_Interp1_Get_public.inc`, note the updating of the local argument `argName`, which is then used in the `catch` block to give a more informative error message. If the translation code (*e.g.*, the call to `Repository::Fetch`) fails, the context of the failure will be made apparent.

We take further advantage of this by moving the validity checks for an argument into the interface description. For example, as part of the function to construct a cubic-spline interpolant we might write:

```
x is number[]
    &IsMonotonic(x)\x values must be in ascending order
    The x-values at which the function value is specified
```

A condition applied to a single argument is placed together with the argument's help, but prefixed with & to distinguish it.[4] The backslash separates the test to be applied in code from the error message to be displayed in case of failure, which in turn doubles as display text in a generated help file. If no such text is supplied, the text of the test itself is used. Our & construct takes a code snippet as its first argument, and an optional human-readable comment as its second after a separating '\' (if the latter is not specified, then a plausible one is generated from the code snippet). The resulting C++ code is:

```
REQUIRE(IsMonotonic(x), "x values must be in ascending order");
```

If the *x*-values are not monotonic, the resulting exception will be caught within the machine-generated wrapper function; both the message and the argument name will be passed to `Excel::Error` to create the return value. Thus the user will likely see a redundant error message in this case, like:

```
"#ERROR -- x values must be in ascending order (at x (input #2))"
```

---

[4]The idea being that the argument has a type *and* a condition.

Often, the argument name is not part of the exception – *e.g.*, because the input was rejected by `Excel::ToDoubleVector`. Note that we set the argument name to `"x (input #2)"`, to provide as much information as possible; user complaints about excessively informative error messages are rare.

In writing validation code, Machinist replaces the special character `$` with the argument name in the generated code, error messages, and help.[5] This allows constructs like

```
end_date is date
    &$ >= start_date\$ cannot precede start date
```

which can be shorter and clearer than the "expanded" version (especially for arguments with multiple constraints).

3.4.3 *Containers*

We will customize the STL `vector` to create our own `Vector_`; see Sec. 4.1 for a fuller discussion.

The standard library `string` is adequate for most purposes, except that it is case-sensitive. Quants can forget that, outside the narrow world of C and C++, case is typically ignored. We can use the standard template library "traits" of `string` to construct a case-insensitive string; this is a straightforward exercise, but we must be careful to make character comparison as efficient as possible (`toupper` is far too slow for this purpose). The best solution appears to be a static data table, which will be replicated in each source file:

```
————————————————— Strings.h —————————————————
     namespace
     {
         // handwritten lookup table -- specifies ordering
         static const unsigned char CI_ORDER[128] =
 5       { 0, 1, 2, 3, 4, 5, 6, 7, 8, 9, 10, 11,
         //...
         84, 85, 86, 87, 88, 89, 90, 97, 98, 99, 100, 101 };
     }
     // traits for case-insensitive string
10   struct ci_traits : char_traits<char>
     {
         typedef char _E;
         static inline unsigned char SortVal(const _E& _X)
         {
15           unsigned char X(_X);
```

---

[5] Granted, this would make generation of Perl code gratuitously difficult.

```
          return (X & 128) | CI_ORDER[X & 127];
      }
      static bool __cdecl eq(const _E& _X, const _E& _Y)
      {
20        return SortVal(_X) == SortVal(_Y);
      }
      static bool __cdecl lt(const _E& _X, const _E& _Y)
      {
          return SortVal(_X) < SortVal(_Y);
25    }
      static const _E* find(const _E* _P, int _N, const _E& _A)
      {
          auto a = SortVal(_A);
          while (_N && SortVal(*_P) != a)
30            ++_P, --_N;
          return _N ? _P : nullptr;
      }
      static int compare(const _E* _P1, const _E* _P2, size_t _N)
      // ...
35 };
```

Of course we will choose the ordering so that CI_ORDER['a'] ==
CI_ORDER['A']. Repetition of this array in (almost) every translation unit
increases the size of our DLLs by around 1% – a substantial impact for so
little functionality, but not intolerable. We have used the fact that characters beyond 127 are non-printing and their ordering need not have any
special properties. We can simplify the logic of ci_traits::eq and lt by
using a 256-character lookup table, but this will double the memory use.

The C++ standard library does not provide a hash function suitable for a
case-insensitive string (hash<string> will produce case-dependent output),
so we must provide our own specialization.

We choose to implement String_ as a class rather than a typedef
of basic_string; this allows it to be forward-declared but forces us to
explictly provide constructors (which simply forward to the basic_string
constructor). This inheritance, like that of Vector_, is safe as long as we
add no data members.

```
——————————— Strings.h ———————————
class String_ : public basic_string<char, ci_traits>
{
    typedef basic_string<char, ci_traits> base_t;
public:
5   String_() {}
    String_(const char* src) : base_t(src) {}
    String_(const base_t& src) : base_t(src) {}
    String_(size_t size, char val) : base_t(size, val) {}
    template<class I_> String_(I_ begin, I_ end)
10       : base_t(begin, end) {}
    // also add an implicit constructor from string
```

```
     String_(const string& src)
     : base_t(*reinterpret_cast<const String_*>(&src)) {}
     // support our swap idiom
15   void Swap(String_* other) { swap(*other); }
   };
```

Unfortunately, the standard library does not recognize any commonality between variants of `basic_string` using different traits. This is a defect in the standard, but I am not aware of any plans to remedy it. For instance, our `String_` class is streamed to an output file exactly as a `std::string` would be; but C++ does not recognize this and provides no implementation of `ostream::operator<<` we can use. Our solution is to provide implementations for our own string class using `reinterpret_cast` to allow it to masquerade as a `string`.

3.5 Machinist

"Machinist" is an open-source code generator.[6] Its operation is highly configurable, but always contains four main phases. First, specified files are scanned for mark-up blocks. Then each block's type is found (*e.g.*, `public` in the example above) and a type-specific parser converts the block's text to a nested tree structure (Machinist's `Info_` class; a property tree, but with each node containing pointers to its parent and the tree's root). For each type, a set of files to write and functions of `Info_` to put in each are specified in the configuration: a type-specific `Emitter_` is constructed implementing these functions, which may be in C++ or in *template files* parsed at run time. Finally, the emitter-generated code is compared with the existing file contents, and changed files are overwritten; this avoids triggering unnecessary recompilation.

The mark-up displayed above is parsed by the dedicated `ParsePublic_` class in Machinist, creating a property tree.[7] The tree could also be entered more directly, with levels of indentation showing the nesting (compare with the mark-up shown at the start of Sec. 3.2):

```
  public Interp1_Get
  help:Interpolate a value at specified abcissas
  input:f
     type:handle
5    subtype:Interp1
```

---

[6]Available at `http://bitbucket.org/hyer/Machinist`.

[7]The proliferation of parsers (for the configuration file, the template files, and each mark-up type), and of the idiosyncratic languages implicit therein, is a weakness of Machinist.

```
      help:The interpolant function
   input:x
      type:number
      dimension:1
10    help:The x-values (abcissas)
   output:y
      type:number
      dimension:1
      help:The interpolated function values at x-values
```

Admittedly, this is more useful for illustrating the code-generation process than for communicating the intent of `Interp1_Get`. It does serve to show the close coupling between parser and emitter, since the two communicate by key strings (`type`, `dimension`, etc.). Machinist's public parser and emitter also support several features not used in this simple example; these include overloaded input types, default input values, insertion of handwritten preprocessing code, and links between help pages.

The template code used to specify emitters is somewhat unappetizing. This is largely necessary, since special characters in many languages must be emitted without the template parser's mistaking them for part of its own grammar; we use a very restricted grammar in which all control sequences are introduced by %.

For example, error-checking code for a public function is created from the template code:[8]

```
──────────────── Public.mgt ────────────────
%:EnforceCondition:
      REQUIRE(%_(), %|{%*[help]{%"()}}{"Validation failed: "`
%"()});` emit stringified condition if help is not available
```

Here %: introduces a function definition, %_ calls a built-in function which emits the top-level contents of an `Info_`, %*[]{} defines an iteration over children, etc. This syntax, while not a joy to view, fully separates the mark-up instructions from the code being emitted. We permit ourselves the luxury of using the un-escaped[9] backquote character to introduce a comment, and as a line-continuation marker.

The interface mark-up for a function may request some customization of the machine-generated code, besides entering an argument list. In particular, a function may be marked as:

---

[8]This is the code for function-level conditions; the template for conditions attached to an argument is more involved.

[9]The escape character is of course %.

- Using *slack* parsing of inputs. This is used, *e.g.*, at the Python interface to allow scalars to be used interchangeably with single-element lists.
- Suppressing output of the wrapper code for some interfaces; for instance, functions which interrogate the object repository are useful only in Excel.
- Being *volatile* in Excel (so that it is always recalculated).

Top-level directives in the mark-up block, such as -`python` or `slack`, will instruct the interface generator to emit (or not) the appropriate code. Other mark-up types, such as enumerations or storable objects, often have similar customizations.

Several conventions we use here, such as the /*IF and -IF delimiters of mark-up blocks and the names of files where generated code is placed, are configuration options in Machinist. The relevant segment of the configuration file is

```
——————————— config.ifc ———————————
<-*.cpp;*.h;*.hpp!MG*
    /*IF
    -IF
public
->MG_#_public.inc
    ExcelSource
->MG_#_public.htm
    HTMLHelp
```

This specifies which files Machinist is to scan, and the delimiters to search for; and also, for each mark-up type, which files to create (with the block name substituted for #) and what functions to emit into them.

This provides us with the tools to easily generate trustworthy interface wrappers, but we still must supply the code components for them to use. These are functions such as `Excel::ToDoubleVector` which implement the foreign-function interface to the target platform; they are not displayed here.

3.6 Exception Messaging

Another duty of an interface is to communicate errors as informatively as possible. Our exception class is responsible for carrying this information out to the interface:

```
——————————— Exceptions.h ———————————
class Exception_ : public runtime_error
{
public:
    Exception_(const char* msg);
    Exception_(const string& msg)
```

```
                : Exception_(msg.c_str()) {}
      Exception_(const String_& msg)
                : Exception_(msg.c_str()) {}
};
```

We do not begrudge the call to `c_str` here, because there is no expectation that construction of an exception should be extremely efficient.

We often wish to attach extra information to an exception, which will be viewed later using `std::exception::what`. Our first thought might be to tack this information onto an already-thrown exception. We could accomplish this with `try` blocks:[10]

```
   // an example of code to be avoided
   try
   {
      // interesting code
 5 }
   catch (Exception_& e)
   {
      e.Append("Fitting instrument with maturity "
           + String::FromDate(pi->Maturity()));
10    throw;
   }
```

This is obviously verbose, but its effects in practice are still worse: what if the information (accessed through the iterator `pi` above) is no longer in scope? To make this approach work, this kind of `try` block must be nested within the iteration, rather than wrapped around it, leading to repeated trying at each level of scope.

To avoid writing such monstrous code, we create a mechanism to append environment information to exceptions.

```
───────────────────── Exceptions.h ─────────────────────
   class XStackInfo_
   {
      const char* name_;
      const void* value_;
 5    enum class Type_ { INT, DBL, CSTR, STR, DATE, DATETIME, VOID }
           type_;
      template<class T_> XStackInfo_(const char*, T_) {}

   public:
10    XStackInfo_(const char* name, const int& val);
      XStackInfo_(const char* name, const double& val);
```

---

[10]Be sure to use `throw;` rather than `throw e;` to rethrow an exception, as the latter can lose type and member information.

```
    XStackInfo_(const char* name, const char* val);
    XStackInfo_(const char* name, const Date_& val);
    XStackInfo_(const char* name, const DateTime_& val);
15  XStackInfo_(const char* name, const String_& val);
    XStackInfo_(const char* msg);    // for VOID type
    string Message() const;
};
```

The `Message` function returns a `string`, rather than our own `String_`, because we need to interact with the `runtime_error` in the `Exception_` class.

The internals of `XStackInfo_` are almost absurdly unsafe. They are type-unsafe, because we store all types in a common `void*`,[11] and hold durable references to ephemeral stack variables. It canot even be made `noncopyable`. However, as we will see shortly, it is constructed only by macros which serve to ensure it will be used safely.

`Message` will be called when an exception is thrown, supplying information about the call stack to yield higher quality messages with minimal extra code. The implementation of the constructors and of `Message` is the obvious one:

```
––––––––––– Exceptions.cpp –––––––––––
XStackInfo_::XStackInfo_(const char* name, const int& i)
:
name_(name), value_(&i), type_(Type_::INT)
{ }
5
string XStackInfo_::Message() const
{
    static const string EQUALS(" = ");
    switch (type_)
10  {
    case Type_::INT:
        return name_ + EQUALS + to_string
            (*(reinterpret_cast<const int*>(value_)));
    // ...
```

To represent the state of our computation, we should have one collection of `XStackInfo_` for each stack – *i.e.*, for each thread. This is exactly what the boost class `thread_specific_ptr` provides.

```
––––––––––– Exceptions.cpp –––––––––––
Vector_<XStackInfo_>& TheStack()
{
    static boost::thread_specific_ptr
        <Vector_<XStackInfo_>> INSTANCE;
5   if (!INSTANCE.get())    // get is thread-specific
```

---

[11]This could be implemented as a union, but little would be changed.

```
      INSTANCE.reset(new Vector_<XStackInfo_>);
   return *INSTANCE;   // so is operator*
}
```

The `thread_specific_ptr` ensures that each thread will see its own separate instance of `TheStack`. We store this as a vector (per thread) to minimize the cost of frequent `push_back` and `pop_back`; this is acceptable because the `XStackInfo_` elements can be freely copied.

Now `Exception_`'s constructor can append all available stack information:

```
─────────────────── Exceptions.cpp ───────────────────
string MsgWithStack(const char* msg)
{
    string retval(msg);
    for (auto si : TheStack())
5       retval += "; " + si.Message();
    return retval;
}


Exception_::Exception_(const char* msg)
10  :
runtime_error(MsgWithStack(msg))
{   }
```

No special code is now required to throw an `Exception_`: its own constructor appends whatever extra information is available. We now turn to the question of how to supply that information.

3.6.1   *Macro Hackery*

Our approach so far requires the construction of `XStackInfo_` objects, which are very light at runtime but quite unsafe and somewhat verbose. Removal from `TheStack` is too important for mere humans, and must be made fully automatic. In C++, this is accomplished with a temporary object (in `namespace Exception`) whose destructor will pop the stack:

```
─────────────────── Exceptions.h ───────────────────
struct StackRegister_
{
    ~StackRegister_() { Exception::PopStack(); }
    // constructor MUST push something on the stack
5   template<class T_> StackRegister_
        (const char* name, const T_& val)
    {
        PushStack(XStackInfo_(name, val));
    }
```

```
10    StackRegister_(const char* msg)
      {
          PushStack(XStackInfo_(msg));
      }
};
```

The code as shown is still not completely safe – it does not deal with the vanishingly unlikely case where PushStack throws an exception. In this case, PopStack could potentially be called one time too many; we can deal with this by checking inside its implementation whether TheStack() is empty, before calling pop_back on it.[12]

The next step is to provide a straightforward way to create these objects. As a first cut, we write:[13]

—————————— *Exceptions.h* ——————————
```
#define XXNOTICE(u, n, v) \
    Exception::StackRegister_ __xsr##u(n, v)
#define XNOTICE(u, n, v) XXNOTICE(u, n, v)
#define NOTICE(n, v) XNOTICE(__COUNTER__, n, v)
```

Now we have a correct implementation of NOTICE, but it is somewhat awkward: it requires the user to supply a variable (by name) and also a variable name. We should change the last line above to

—————————— *Exceptions.h* ——————————
```
#define NOTICE2(n, v) XNOTICE(__COUNTER__, n, v)
#define NOTICE(x) NOTICE2(#x, x)
//#define NOTICE(x) XNOTICE(x, #x, x)    // also works
```

Now invoking NOTICE(var) does just what the name implies: it makes a note of var's value, and appends it to any error messages sent to the user (from the scope of the NOTICE). If we want to supply a name in the message different from that used in the code (*e.g.*, if the thing being NOTICEd is inside a nested struct), we can fall back on NOTICE2.

We may also wish to append simple location information, *e.g.*, with code like NOTE("Calibrating local vols");. This can be accomplished using the infrastructure of NOTICE; the VOID type in XStackInfo_ exists for this purpose.

—————————————————————————————

[12]For errors not caused by user inputs, such as out-of-memory or connection failures, the extra stack information is not useful; in such a case it may be safer to throw a std::exception.

[13]The intermediate XNOTICE macro is necessary to force __COUNTER__ to be evaluated once, outside XXNOTICE. Since the C preprocessor has neither a gensym nor a let construct, almost any macro worth writing requires these contortions.

There is a potential runtime hazard in `NOTICE`: constructing the `StackRegister_` object may trigger an implicit conversion, causing construction of a temporary object. This is another reason not to provide implicit conversions in general; but we can detect it by adding a catch-all constructor to `XStackInfo_` which will generate a compile-time error:

```
                        ─── Exceptions.h ───
// in XStackInfo_ private section
   template<class T_> XStackInfo_(const char*, T_) {}
```

The compiler will prefer the template constructor to an implicit conversion to match a more specific constructor; but since the former is `private`, the attempt to call it is a compile-time protection error.

This exact type matching protects us from type errors in `XStackInfo_`; as long as we capture only C++ variables, we will also be protected from scope errors. Using just this macro, it is difficult for a coder to lift an ephemeral object out of its stack scope; but it is still possible to trick `NOTICE` with code like:

```
   NOTICE(p->m);
   delete p;
```

A discipline of `const` objects and smart pointers limits the scope of this issue. If more control is needed, we can scan the codebase with a regular expression search to ensure that the argument to `NOTICE` is always a single C++ identifier.

We also provide the simple macros:

```
                        ─── Exceptions.h ───
#define THROW(msg) throw Exception_(msg);
#define REQUIRE(cond, msg) if (cond); else THROW(msg);
```

These reduce the disruption to the flow of code which is introduced by error-checking with explicit `if` statements. The construction of the message for `REQUIRE` might have a noticeable run-time cost (*e.g.*, it might contain string arithmetic or number-to-string conversions); thus `REQUIRE` is a macro, rather than an inline function, so `msg` will not be evaluated unnecessarily.

3.7 Environment

Despite our preference for pure functions, the problems of derivatives risk management are ineluctably environment-rich. Given a trade as described by its confirmation, a model representing a distribution of future states of the world of prices, and a history representing past states, there are still other facts which must be specified. For instance:

- Has the Bermudan option to terminate this trade already been exercised?
- Has a given reference credit defaulted according to the terms of this trade?
- What bonds exist which can be delivered against this futures contract?

To address questions like these, we posit a C++ class `Environment_` which will be input to functions which can be affected by it.[14]

Of course, we could avoid passing the environment as an explicit input, visually simplifying the interface to our functions, and then internally marshal facts about the world as they are needed. Since much of this information is not rapidly changing and not subject to user input, this is a temptation to which many quants have yielded.

The eventual price is high. Rather than create a function, which can be tested in isolation against a variety of inputs, we have created a procedure which takes actions of its own. During testing and debugging, such code seems alive – and hostile. It can still be tested, but at the cost of creating a controlled reconstruction (a *mock*) of whatever method supplies the external environment.

In short, we can build testability into our code; or, once having discarded it, we can try extreme measures to recover. By making the environmental access more explicit, we bring it more under control.

The proper design of `Environment_` is an interesting illustration of our programming principles. An improperly designed environment would become a compile-time chokepoint, bringing together many types of functionality under a common interface which would in turn be seen by any code needing any type of environment (see Sec. 2.4.1). For example, multi-credit trades would be forced to "know about" deliverable bonds for exchange-traded futures.

To avoid this, we will write in C++:

```
———————————————— Environment.h ————————————————
namespace Environment
{
    struct Entry_ : noncopyable
    {
```

---

[14]More stable facts about the world, such as holiday calendars, are treated differently; see Secs. 2.4.1 and 8.3.

```
5        virtual ~Entry_();
     };
 }

 // inefficient implementation; we will do better below
10 class Environment_ : noncopyable
 {
 public:
     typedef Environment::Entry_ Entry_;
     Vector_<Handle_<Entry_>> vals_;
15 };
```

A `Handle_` is a shared-pointer-to-`const`; this is the only widely acceptable use of shared pointers, since a shared pointer to mutable data gives referential transparency to none of its owners. (Later we will see a few specialized uses for shared pointers to non-`const` data, but they are rare; the loss of data integrity is a great drawback. Often the only code which can safely use such a pointer is its creator and *de facto* owner, so the pointer is not meaningfully shared.)

The end result is an environment that can be passed to functions which themselves have no idea of the multitudinous forms which the environment may contain, because `Environment_` does not know them itself. This is a precondition of extensible design.

Many functions will require a `const Environment_& env` as part of their declaration.

```
────────────────── Environment.h ──────────────────
#define _ENV const Environment_* _env
```

`_ENV` is always the first rather than the last argument: otherwise its presence would interact with changes to function signatures.

We pass the environment as a pointer, rather than a reference; this lets us create a new environment which augments the input with more information, then repoint `_env` to that environment.

The presence of `_ENV` in a function signature is a warning that the function is not "pure" – it does not map inputs to outputs in a unique way. Two programming guidelines will help us create more comprehensible and testable code:

- *Purity*: an input `_ENV` should be used as soon as possible, and the results passed down to pure functions which do the computational work.
- *Honesty*: a function which needs environment should request `_ENV` as input, with documentation as to what is needed, so that environment is populated at a high level (ideally at the public interface).

By propagating the need for environment up but not down, we increase the portion of our library which is purely functional.

3.7.1 Fast-Path Optimization

The above implementation of `Environment_` is simple to read, and efficient when an exception is thrown, but imposes excessive overhead in the non-exceptional case. Consider the code to supply access to a cached query for a set of deliverable bonds. It must look something like:

```
Handle_<Entry_> bdb(new BondAccess_(asof_date);
Enviroment_ augmented(*_env);
augmented.vals.push_back(bdb);
_env = &augmented;
```

We have created a new environment locally, containing the copy the old contents plus an addtional `Handle_<Environment::Entry_>`; both the `Handle_` and the `Vector_` of entries in the new environment require a heap allocation. To avoid this inefficiency, we need to re-implement `Environment_`:

```
                        ———— Environment.h ————
class Environment_ : noncopyable
{
public:
    virtual ~Environment_();

 5
    typedef Environment::Entry_ Entry_;
    struct IterImp_ : noncopyable
    {
        virtual ~IterImp_();
10      virtual bool Valid() const = 0;
        virtual IterImp_* Next() const = 0;
        virtual const Entry_& operator*() const = 0;
    };
    struct Iterator_
15  {
        Handle_<IterImp_> imp_;
        Iterator_(IterImp_* orphan) : imp_(orphan) {}
        bool IsValid() const { return imp_.get() != 0; }
        void operator++();
20      const Entry_& operator*() const { return **imp_; }
    };

    virtual IterImp_* XBegin() const = 0;
    Iterator_ Begin() const { return Iterator_(XBegin()); }
25 };
```

This more efficient `Environment_` does not store all its entries in one place; but it provides an `Iterator_` which can be used to access all the entries there are. We supply convenience functions in `namespace Environment` to scan the entries:

```
——————————————————— Environment.h ———————————————————
     template<typename F_> void Iterate
        (const Environment_* env, F_& func)
     {
        if (env)
 5          for (auto pe = env->Begin(); pe.IsValid(); ++pe)
               func(*pe);
     }


     template<typename F_> auto Find
10      (const Environment_* env, F_& func)
        -> decltype(func(*env->Begin()))
     {
        if (env)
        {
15         for (auto pe = env->Begin(); pe.IsValid(); ++pe)
               if (auto ret = func(*pe))
                   return ret;
        }
        return 0;
20   }
```

Since by far the most common scan is for entries of a given type, we provide special-purpose functions for that as well:[15]

```
——————————————————— Environment.h ———————————————————
     template<typename T_> Vector_<const T_*> Collect
        (const Environment_* env)
     {
        Vector_<const T_*> retval;
 5      Iterate(env, [&](const Handle_<Entry_>& h)
               {if (dyn_ptr(t, const T_*, h.get()))
                   retval.push_back(t); });
        return retval;
     }

10
     template<typename T_> const T_* Find
        (const Environment_* env)
     {
        return Find(env, [](const Entry_& e)
15             {return dynamic_cast<const T_*>(&e); });
     }
```

---

[15] These look a lot better with longer lines.

The overload of **Find**, so that the one-argument version defaults to finding by type, is quite natural in practice because the type being found must be explicitly specified: *e.g.*, **Find<InstrumentDB_>(env)**.

Our first version of **Environment_** (now renamed to **Environment::Base_**) can easily provide a concrete implementation of the new interface:

```
_____ Environment.h _____
class Base_ : public Environment_
{
    Vector_<Handle_<Entry_>> vals_;
    struct MyIter_ : IterImp_
    {
        const Vector_<Handle_<Entry_>>& all_;
        Vector_<Handle_<Entry_>>::const_iterator me_;
        MyIter_(all_, me_) {}
        bool Valid() const override
        {
            return me_ != all_.end();
        }
        IterImp_* Next() const override
        {
            return new MyIter_(all_, ::Next(me_));
        }
        const Entry_& operator*() const override
        {
            return **me_;
        }
    };
public:
    Base_(vals_ = Vector_<Handle_<Entry_>>()) {}
    MyIter_* XBegin() const override
    {
        return new MyIter_(vals_, vals_.begin());
    }
};
```

Base_ will also serve to create empty environments.

To easily add information to the environment, we need a derived class which prepends information to an existing environment. We place this in **namespace Environment**; the **X** in its name reflects the fact that it depends on local references (as discussed in Sec. 3.6) and thus may not be passed out of its local scope.

```
─────────────────────── Environment.h ───────────────────────
 class XDecorated_ : public Environment_
 {
    const Environment_*& theEnv_;    // of our scope
    const Environment_* parent_;
5   const Entry_& val_;

    struct I1_ : IterImp_
    {
       const Environment_* parent_;
10     const Entry_& val_;
       I1_(parent_, val_) {}
       bool Valid() const { return true; }
       IterImp_* Next() const
       {
15        return parent_ ? parent_->XBegin() : 0;
       }
       const Entry_& operator*() const { return val_; }
    };
 public:
20  XDecorated_(theEnv_, val_)
    {
       parent_ = theEnv_;
       theEnv_ = this;
    }
25  ~XDecorated_() { theEnv_ = parent_; }

    IterImp_* XBegin() const
    {
       return new I1_(parent_, val_);
30  }
 };
```

Now `XDecorated_` performs two roles: it provides the augmented environment, and manages the `env` pointer of the scope in which it is created.
Notice that creation of an enhanced local environment is very similar to
the `NOTICE`ing of stack info in Sec. 3.6.1; instead of popping the thread's
environment stack, we reset the environment pointer to its previous value.

We can embed this functionality in macros similar to `NOTICE`:

```
─────────────────────── Environment.h ───────────────────────
 #define XX_ENV_ADD(u, e) \
        Environment::XDecorated_ __xee##u(_env, e)
 #define X_ENV_ADD(u, e) XX_ENV_ADD(u, e)
 #define ENV_ADD(e) X_ENV_ADD(__COUNTER__, e)
5 #define ENV_SEED(e) _ENV = nullptr; ENV_ADD(e)
```

These macros should be the only way in which `Environment::XDecorated_` is used; we can verify this by global text search of the code base. Unfortunately, the environment entries might not be simple stack variables, so constructions like `ENV_ADD(*db)` will inevitably occur; we cannot repeat the trick of searching for arguments other than identifiers. We are also relying on developers to be parsimonious in grabbing environment resources, preferring functions that receive those resources in their input `_env`, or receive necessary information as an explicit input. Nonetheless, these macros provide a simple and reasonably safe way to add environment where it is required.

3.7.2 *Repository Access*

The repository of stored handles is obviously part of the environment: thus our preferred way to access it will be through the `Environment_`. Since the repository only exists at the interface level, the header providing this functionality is also at the interface level.

```
————————————— _Repository.h —————————————
     #include "Environment.h"
     #include "Strings.h"
     class Storable_;

5    class ObjectAccess_ : public Environment::Entry_
     {
         String_ AddBase
             (const Handle_<Storable_>& object,
              const RepositoryErase_& erase)
10       const;

     public:
         Handle_<Storable_> Fetch
             (const String_& tag, bool quiet = false)
15       const;

         int Size() const;
         Handle_<Storable_> LowerBound(const String_& part_name) const;
         Vector_<Handle_<Storable_>> Find(const String_& pattern) const;
20       int Erase(const String_& pattern) const;
         bool Erase(const Storable_& object) const;

         template<class T_> String_ Add
             (const Handle_<T_>& object,
25            const RepositoryErase_& erase)
         const
         {
             return AddBase(handle_cast<Storable_>(object), erase);
         }
30   };
```

This class defines the interface to the repository. Note that we can `Erase` a single object, or all objects whose repository keys match a pattern (which will be interpreted as a `std::regex`). The strange-looking `LowerBound` function provides efficient support for storage of fixings and global dates; see Sec. 9.2.

All these functions simply forward to corresponding functions in a local namespace in `_Repository.cpp`; see Sec. 5.6. They use no member data, but are declared `const` rather than `static` because they should only work if an actual instance of `ObjectAccess_` exists. The effect is that there is a single in-process repository, to which every instance of `ObjectAccess_` provides an interface.

Since `ObjectAccess_` is default-constructed – it is just a statement of our intent to interact with the environment in a particular way – we can help prevent unsafe constructions like `ENV_SEED(ObjectAccess_())` by providing zero-argument forms of `ENV_ADD` and `ENV_SEED`:

```
─────────────────────── Environment.h ───────────────────────
#define XX_ENV_INST(u, t) t __ei##u; ENV_ADD(__ei##u)
#define X_ENV_INST(u, t) XX_ENV_INST(u, t)
#define ENV_ADD_TYPE(t) X_ENV_INST(__COUNTER__, t)
#define ENV_SEED_TYPE(t) _ENV = nullptr; ENV_ADD_TYPE(t)
```

We have the option to enforce this as the only form of repository access, but it involves a complicated web of `friends` (for example, `ObjectAccess_` must be given a private constructor, otherwise any function can simply make one, even without any `Environment_`). My own experience is that such policing is not necessary: it is enough to provide a simple way to do the right thing.

As a rule, the coder will not interact directly with these functions at all; the repository should be accessed only from the Excel-specific machine-generated code. Thus we include in each Excel wrapper:

```
    ENV_SEED_TYPE(ObjectAccess_);
```

This ensures the accessibility of the repository from the environment, which is then passed where it is needed:

```
─────────────────────── MG_Interp1_Get_public.inc ───────────────────────
    // ...
    auto f = Excel::ToHandle<Interp1_>(_env, xl_f);
```

The implementation of `ToHandle`, rather than appeal directly to the repository, will call the utility function:

```
_____ _Excel.cpp _____
Handle_<Storable_> Excel::ToHandleBase
    (_ENV, const OPER_* src, bool optional)
{
    const String_ tag = ToString(src, optional);
5   if (tag.empty())
    {
        REQUIRE(optional, "Missing input handle");
        return Handle_<Storable_>();
    }
10  auto repo = Environment::Find<ObjectAccess_>(_env);
    assert(repo);
    return repo->Fetch(tag);
}
```

Note that the call to `repo->Fetch` does not allow the result to be optional – an optional handle may be omitted altogether, but we must not silently ignore an invalid handle string.

3.8 Enumerated Types

We have mentioned the need for enumerated types, constructible from strings. These can be machine-generated from a high-level description. For example, our mark-up might read:

```
_____ OptionType.h _____
/*IF-------------------------------------------------------
enumeration OptionType
    Call/put flag
switchable
5 alternative CALL C
alternative PUT P
alternative STRADDLE V C+P
method double Payout(double spot, double strike) const;
method OptionType_ Opposite() const;
10 -IF------------------------------------------------------*/
```

This is sufficient information[16] to define a `class OptionType_` which can be constructed from or converted to a string:

---

[16]Machinist supports help strings for each `alternative`, but defining a call seems excessive.

```
─────────────────────── MG_OptionType_enum.h ───────────────────────
   class OptionType_
   {
   public:
      enum class Value_ : char
5     {
         _NOT_SET=-1,
         CALL,
         PUT,
         STRADDLE,
10        _N_VALUES
      } val_;

      OptionType_(val) {assert(val < Value_::_N_VALUES;}
   private:
15     friend bool operator==(const OptionType_& lhs,
            const OptionType_& rhs) {return lhs.val_ == rhs.val_;}
      friend struct ReadStringOptionType_;
      friend Vector_<OptionType_> OptionTypeListAll();
      friend bool operator<(const OptionType_& lhs,
20        const OptionType_& rhs) {return lhs.val_ < rhs.val_;}
   public:
      explicit OptionType_(const String_& src);
      const char* String() const;
      Value_ Switch() const {return val_;}
25     // idiosyncratic (hand-written) members:
      double Payout(double spot, double strike) const;
      OptionType_ Opposite() const;
      OptionType_() : val_(Value_::_NOT_SET) {};
   };
```

Machinist emits this together with boilerplate code such as operator=!,
and operator== for both class-class and class-enum comparisons (to support type == OptionType::Type_::STRADDLE without needing to construct a temporary variable). In MG_OptionType_enum.inc we will
machine-generate any out-of-line implementations supporting this interface.

The first text of each ALTERNATIVE is the *canonical name* for that value,
which is part of the enumerated Type_ and will be the output of a conversion
to string; additional valid inputs with the same meaning (like "V" to mean
STRADDLE) may follow. This convention requires that the canonical name
be a valid C++ identifier. Aliases such as C+P will be used only in string
recognition, and have no such limitation.

Our parsing of input strings should suppress whitespace; thus

```
OptionType_("c + p").String()
```

evaluates to "STRADDLE" (strings are case-insensitive, as always). We prefer to suppress underscores as well, so that ACT360 and ACT_360 will be considered equivalent.

The method keyword introduces a new member function (to be included in the class definition).[17] Since our entire library is in C++, we can define these members by giving their C++ function signatures; to support multiple languages, we would have to introduce separate keywords like method:C++ for each language,[18] or else describe methods in a more abstract way (as we do for public functions) from which we could generate code for each target language.

The mark-up declared this type switchable, *i.e.*, interconvertible with its enum type; thus we define the Type_ in the class's public section and add the conversion members. Omitting the switchable keyword would make the enum private to the class.

Because a switchable enumeration can be manipulated directly through its enumerated value, we could consider supporting a public interface based on this value; this would permit coders outside our library to maintain their own option type enumeration, which would be converted to an int to pass through our public interface. This approach introduces two gratuitous diffculties. First, it cannot work for non-switchable types, for which we can only accept string inputs. Second, it requires the enumeration to be maintained synchronously in two separate codebases; thus we will be forced, at the very least, to synchronize releases and ensure that they are never mismatched. It is far better to treat all public enumerations in the more general way, and pass human-readable strings at the interface.

Machine-generated enum classes are our preferred method of introducing any control switch. For example, in Sec. 3.7.2 we introduced a type RepositoryErase_, which can be defined in mark-up:[19]

```
                        _Repository.h _
/*IF-------------------------------------------------
enumeration RepositoryErase
help:Controls what is erased when a new tag is added
help:to the repository
switchable:1
alternative:NONE
    help:Erase nothing, just add
alternative:NAME_NONEMPTY
```

---

[17]We can never introduce new member *data*, which would defeat our purpose.
[18]More precisely, for each language with closed class definitions.
[19]In this example we have used the mark-up to create the property tree directly, but Machinist also supports a parser for enumeration types.

```
       default:1
10     help:Erase object of same type and name, iff name
       help:is nonempty
    alternative:NAME
       help:Erase object of same type and name
    alternative:TYPE
15     help:Erase all objects of the same type
    -IF-------------------------------------------------*/
```

See Sec. 8.1.4 for other examples. There is no reason *not* to use this method
for any enumerated switch: it ensures that enumerations will be docu-
mented, efficiently constructed from strings, and in a standardized form.

We will return to enumerated types in Sec. 8.1, and discuss run-time
extension of the set of possible values.

Chapter 4

Vector and Matrix Computations

Linear algebra, especially matrix decompositions, and efficient computation with arrays are constant necessities. By carefully structuring the interfaces of these routines, we can write more flexible and safer code.

4.1 Customizing Vectors

The standard library `std::vector` is a fairly good fit for our needs: in particular, its iteration and subscripting are optimal. We would prefer, however, to have somewhat more numerical support.

One important decision is whether to use the same container (or container template) for general-purpose data storage and for processor-intensive arithmetic. By separating the two, we could highly optimize the numerical containers without interfering with the ease of use of storage containers. Our judgement is that the price to be paid in increased code complexity is too steep; we will focus instead on improving `vector`.

We will remove as well as adding functionality: in particular, the two-argument form of `std::vector::resize` is confusing even to experienced programmers.

Our implementation will follow the standard library implementation of the adaptors `stack` and `queue` – these adapt the interface of another object but not its data members, making private inheritance safe. Since the most common element type is `double`, we make that the default, writing simply `Vector_<>` when we need a vector in the mathematical sense.

The definition of `Vector_` uses `using` declarations to bring `std::vector` functionality into its interface; functions like the two-argument `resize` are suppressed simply by omitting this declaration. We add member operators `+=` and `-=` (adding/subtracting another vector, or a scalar), and a single

47

operator*= to rescale. (Inner product is a nonmember algorithm, not an
operator; see Sec. 4.2.)

```
———————————————— Vectors.h ————————————————
template<class E_> class Vector_ : private vector<E_>
{
public:
    Vector_() : vector<E_>() {}
5   Vector_(int size) : vector<E_>(size) {}
    // ...

    void Swap(Vector_<E_>* other) { swap(*other); }
    void Fill(const E_& val) {fill(begin(), end(), val);}
10  void Resize(int new_size) { resize(new_size, E_()); }

    template<class T_> void operator*=(const T_& scale)
    // ...
    template<class C_> void Append(const C_& other)
15  { insert(end(), other.begin(), other.end()); }

    // reveal std::vector types
    using vector<E_>::iterator;
    // ...
20  // reveal std::vector functionality
    using vector<E_>::begin;
    //...
    int size() const {return static_cast<int>(vector<E_>::size());}
};
```

Here we show only a representative cross-section of the functionality of
Vector_. C++ insists that the default for a template argument be sup-
plied everywhere or nowhere; thus, in **Platform.h**, we both forward-declare
Vector_ and supply the default element type.

```
———————————————— Platform.h ————————————————
template<class E_ = double> class Vector_;
```

4.2   Algorithms

The C++ standard library algorithms are deliberately designed for maximum generality, at the expense of making them unpleasantly verbose in
many contexts. For example, an in-place **transform** requires the container
being operated on to be named thrice – since the name may be nontrivial,
such as a nested sub-object or the result of a computation, this leads to
cluttered and error-prone code.

Thus we create many container-level algorithms, echoing those defined at iterator level in the STL. When possible, we name these identically to the underlying algorithm except for capitalization; this causes no confusion since the purpose of the algorithm is unchanged.[1]

As an example, here are the implementations of overloaded forms of Transform:

```
——————————— Algorithms.h ———————————
template<class CS_, class OP_, class CD_>
void Transform(const CS_& src, OP_ op, CD_* dst)
{
    assert(dst && src.size() == dst->size());
5    transform(src.begin(), src.end(), dst->begin(), op);
}

template<class CS1_, class CS2_, class OP_, class CD_>
void Transform(const CS1_& s1, const CS2_& s2, OP_ op, CD_* dst)
10   {
    assert(dst && s1.size() == dst->size()
        && s2.size() == dst->size());
    transform(s1.begin(), s1.end(), s2.begin(), dst->begin(), op);
}

15   template<class C_, class OP_>
void Transform(C_* to_change, OP_ op)
{
    assert(to_change != nullptr);
20   transform(to_change->begin(), to_change->end(),
        to_change->begin(), op);
}

template<class C_, class CI_, class OP_>
25   void Transform(C_* to_change, const CI_& other, OP_ op)
{
    assert(to_change != nullptr);
    transform(to_change->begin(), to_change->end(),
        other.begin(), to_change->begin(), op);
30   }
```

Unless efficiency is at an absolute premium, we prefer a more functional interface which creates and populates a vector in a single step:[2]

```
——————————— Algorithms.h ———————————
template <typename C, typename OP> auto Apply
    (OP op, const C& src)
->typename vector_of<decltype(op(*src.begin()))>::type
{
```

---

[1] In fact, names that look the same to the human but different to the compiler – allowing disambiguation in some cases – are ideal here.

[2] This function is called map in most functional languages.

```
5    vector_of<decltype(op(*src.begin()))>::type retval
         (src.size());
     Transform(src, op, &retval);
     return retval;
  }

10 template<class C1_, class C2_, class OP_> auto Apply
     (OP_ op, const C1_& src1, const C2_& src2)
   ->typename vector_of
     <decltype(op(*src1.begin(), *src2.begin()))>::type
15 {
     vector_of<decltype(op(*src1.begin(), *src2.begin()))>
         ::type retval(src1.size());
     Transform(src1, src2, op, &retval);
     return retval;
20 }
```

We extend this notation to other STL algorithms, *e.g.* Copy,
LowerBound, BinarySearch, InnerProduct, Accumulate.[3] For the STL
unique algorithm, we sort the input as well:

─────────────── ***Algorithms.h*** ───────────────
```
  template<class C_, class LT_>
  C_ Unique(const C_& src, const LT_& less)
  {
      C_ ret(src);
5     sort(ret.begin(), ret.end(), less);
      ret.erase(unique(ret.begin(), ret.end()), ret.end());
      return ret;
  }
  template<class C_> C_ Unique(const C_& src)
10 {
      return Unique(src, less<C_::value_type>());
  }
```

Since we rarely use std::list, we do not bother to specialize the first
version of Unique.

Other useful functions, gleaned from those commonly used in pure functional programming, include Zip, which generates a vector of pairs from a
pair of vectors, and Unzip; Filter, which encapsulates STL's remove_if;
and Keys, which gets the keys of a map (or other associative container).

──────────────────────

[3]These latter two are placed in Numerics.h, following the standard library convention.

4.2.1 *Join*

In addition to manipulating vectors, we often want to concisely construct them from individual elements. This is made possibly by a few template functions:[4]

```
_____ Vectors.h _____
template<class E_> Vector_<E_> Join
    (const E_& e1, const E_& e2)
{
    Vector_<E_> retval(1, e1);
    retval.push_back(e2);
    return retval;
}
Vector_<String_> Join
    (const char* e1, const char* e2)
{
    return Join(String_(e1), String_(e2));
}
template<class E_> Vector_<E_> Join
    (const E_& head, Vector_<E_>& tail)
{
    Vector_<E_> retval(1, head);
    Append(&retval, tail);
    return retval;
}
template<class E_> Vector_<E_> Join
    (const Vector_<E_>& sofar, const E_& next)
{
    Vector_<E_> retval(sofar);
    retval.push_back(next);
    return retval;
}
```

For example, `Join(Join("TRADE", "MODEL"), "BUCKET")` constructs a 3-element vector; the template specialization of `Join` for `const char*` is necessary to ensure that it creates a `Vector_<String_>`.

In C++11 a vector can be constructed directly from built-in elements with known values:

```
vector<int> v = {1, 2, 3};
```

This language advance replaces many uses of `Join`.

We use another version of `Join` to merge two containers:

```
_____ Algorithms.h _____
template<class C1_, class C2_> C1_ Join
    (const C1_& c1, const C2_& c2)
```

---
[4]Yes, this is just cons.

```
   {
       C1_ retval = c1;
5      Append(&retval, c2);
       return retval;
   }
```

Here `Append` is itself a container-level operation, with a specialization for `Vector_` which does not support the possibly-inefficient `insert`:

```
─────────────── Algorithms.h ───────────────
template<class C1_, class C2_> void Append(C1_* c1, const C2_& c2)
{
    c1->insert(c1->end(), c2.begin(), c2.end());
}
5 template<class E1_, class C2_> void Append
    (Vector_<E1_>* c1, const C2_& c2)
{
    c1->Append(c2);
}
```

4.3   Matrices and Square Matrices

4.3.1   *Internal Layout*

We choose to use row-major contiguous storage for our `Matrix_` class: this reflects our emphasis on maximal efficiency of access and minimal allocation calls, at the expense of inefficient resizing (since the values must be shuffled around whenever the number of columns changes). We make several requirements of `Matrix_`:

- It must produce ephemeral `Row_` and `Column_` sub-objects which view or manipulate its data;
- These must have `ConstRow_` and `ConstColumn_` relatives for viewing only;
- It must support an ephemeral `SubMatrix_` which sees part of its data, and which also produces `Row_`s and `Column_`s.

As a result, our `Matrix_` implementation is large (over 150 lines of code), and we show only fragments of it here.

```
─────────────── Matrix.h ───────────────
template<class E_> class Matrix_
{
    Vector_<E_> vals_;
    int cols_;
5   typedef typename Vector_<E_>::iterator I_;
```

```
     Vector_<I_> hooks_;

     void SetHooks(int from = 0)
     {
10       for (int ii = from; ii < hooks_.size(); ++ii)
             hooks_[ii] = vals_.begin() + ii * cols_;
     }
 public:

15   //...
     // Slices -- ephemeral containers of rows or columns
     class ConstRow_
     {
     protected:
20       I_ begin_, end_; // non-const to support Row_ too
     public:
         typedef E_ value_type;
           typedef typename Vector_<E_>::const_iterator
                 const_iterator;
25       // construct from begin/end or from begin/size
         ConstRow_(begin_, end_) {}
         ConstRow_(begin_, int size) : end_(begin + size) {}

         const_iterator begin() const { return begin_; }
30       const_iterator end() const { return end_; }
         size_t size() const { return AsInt(end_ - begin_); }
         const E_& operator[](int col) const
         { return *(begin_ + col); }
     };
35   ConstRow_ Row(int i_row) const
     { return ConstRow_(hooks_[i_row], cols_); }
     ConstRow_ operator[](int i_row) const
     { return Row(i_row); }     // C-style access

40   struct Row_ : ConstRow_
     {
         // inherit data members and const_iterator
         typedef I_ iterator;
     //...
```

4.3.2 *Pasting and Formatting*

This row-major storage scheme means that appending a second matrix by adding extra rows ("append to bottom") is sometimes efficient, while adding extra columns ("append to right") is not. Thus only the former is supported by a special-purpose function:

```
                          ──── Matrix.h ────────────────────
   namespace Matrix
   {
      template<class E_> void Append
         (Matrix_<E_>* above, const Matrix_<E_>& below);
 5 }
```

For all other combinations, we write a generic function to glue matrices together in a user-specified format. In our experience, this functionality is only used at the user-interface level, so we provide it only for matrices of `Cell_`.

```
                     ──── MatrixUtils.h ────────────────────
   namespace Matrix
   {
      Matrix_<Cell_> Format
         (const Vector_<Matrix_<Cell_>*>& args,
 5       const String_& format);
   }
```

This function takes a vector-of-pointers, rather than a vector, as input to avoid unnecessary copying; it does not own or memory manage the pointers. As a rule, when merging only a few matrices, we will just `Join` their addresses to obtain the vector of pointers.

The format string defines the operations we might perform. For our example code, we use a simple syntax using notation which will be familiar to Excel users.

- The numbers 1, 2, ... refer to the elements of the vector in sequence. Note the use of 1-offset.
- The number 0 refers to a 1x1 matrix of an empty `Cell_`; this can be used to insert spacers.
- The suffix `T` after a term means the transpose of that term.
- The suffix `I` after a term means the inverse (180° rotation) of that term.
- The suffix `*` after a term maps that term's contents into a single row vector (beginning with the first row).
- The operators ",", and ";" mean side-by-side and top-to-bottom join, respectively (as in Microsoft Excel). These are top-justified and left-justified, respectively.
- The operators "." and ":" are bottom-justified and right-justified variants.
- Parentheses form groups which are then manipulated as units.

This enables formats like `1,0,((2,3,6,7)T:(4I,5))` which join several inputs into a highly customized output, without our writing a lot of equally customized code.

The implementation requires a simple parser for the format string, and then a set of output builders to create the output value based on the parsed instructions. If the element type is templated, this functionality should be kept in a single source file, to which non-template functions like `Format` is are the only interface; then the builders will be instantiated only once each, avoiding substantial code bloat.

Formatted merges can also be used by `Excel::Retval_` (see Sec. 3.2) to let users (or mark-up) control the displayed result of a function with several outputs.

4.4 Matrix Multiplication

With row-major matrix storage, left-multiplication of a vector by a matrix (*i.e.*, Mv) is extremely efficient.

```
——————————— MatrixArithmetic.cpp ———————————
   void MultiplyAliasFree
      (const Matrix_<>& left,
       const Vector_<>& right,
       Vector_<>* result)
 5 {
      assert(result != &right);
      result->Resize(left.Rows());
      for (int ir = 0; ir < left.Rows(); ++ir)
         (*result)[ir] = InnerProduct(left.Row(ir), right);
10 }
```

The one complication arises from the possibility of aliasing, which is easily detected because `vectors` (and our `Vector_s`) cannot overlap.

```
——————————— MatrixArithmetic.cpp ———————————
   void Matrix::Multiply
      (const Matrix_<>& left,
       const Vector_<>& right,
       Vector_<>* result)
 5 {
      assert(left.Cols() == right.size());
      if (result == &right)    // aliased
         MultiplyAliasFree(left, Vector_<>(right), result);
      else
10       MultiplyAliasFree(left, right, result);
   }
```

Our version of `MultiplyAliasFree` is a template which can also be used for `Matrix_::SubMatrix_` objects. We do not choose to support an `operator*` for multiplication; in our view, the resulting brevity comes at too high a cost in clarity. This would change if matrix multiplication were a constantly recurring task for us; it is not.

For right-multiplication (*i.e.*, $v^T M$) we can either echo the above implementation exchanging columns and rows in the matrix access, or else resort to a more complex implementation that avoids the slower column iterators. This choice is really a matter of taste; empirically, right-multiplication by nonsymmetric matrices is not used in any performance hotspots, so the observed performance difference is small.

Matrix-matrix multiplication raises the same issues again: as well as checking whether the output overwrites the left- or right-hand side, we must also handle the case where all three are the same matrix.[5]

We need one more function: the *weighted inner product*, $v^T W x$.

```
─────────────── MatrixArithmetic.cpp ───────────────
double Matrix::WeightedInnerProduct
    (const Vector_<>& left,
     const Matrix_<>& w,
     const Vector_<>& right)
{
    assert(left.size() == w.Rows());
    assert(right.size() == w.Cols());
    double retval = 0.0;
    for (int ir = 0; ir < w.Rows(); ++ir)
        retval += left[ir] * InnerProduct(w.Row(ir), right);
    return retval;
}
```

4.4.1 *Inheritance and Substitutability*

A well-known puzzle of object-oriented programming is to ask, "Is a square a rectangle?" That is, can `Square_` be a subclass derived from `Rectangle_`? Perhaps surprisingly, the answer is often no: a method like `Rectangle_::Resize(double dx, double dy)` cannot be supported by any plausible `Square_`. Similarly, any matrix that supports resizing cannot be a base class for a square matrix.

The solution to this conundrum is in the distinction between `const` and mutable objects: a `const Square_` certainly *is-a* `const Rectangle_`. In

---

[5]I am not aware of having written any code which squares a matrix in-place; but I prefer not to find out by blowing it up.

C++, we can express this by implementing `Square_` using `Rectangle_` `imp_` as the sole member datum, and providing an implicit conversion operator:

```
operator const Rectangle_&() const {return imp_;}
```

At this point a `const Square_` can be used anywhere a `const Rectangle_` can. We use exactly this method to ensure that a square matrix, when `const`, is a matrix.[6] This makes it usable in any function expecting a `Matrix_`, but we must still add forwarding functions to give `SquareMatrix_` its own interface.

```
───────── SquareMatrix.h ─────────
template<class E_> class SquareMatrix_
{
    Matrix_<E_> val_;
public:
    SquareMatrix_(int size) : val_(size, size) {}
    void Resize(int size) { val_.Resize(size, size); }

    operator const Matrix_<E_>&() const { return val_; }
    double& operator()(int i, int j) {return val_(i, j);}
    const double& operator()(int i, int j) const
    { return val_(i, j); }
    // ...
```

4.5 Decompositions (Square)

We will focus here on the interface, rather than the implementation, of matrix decompositions. For most purposes we decompose only square matrices, and our interface reflects this.

```
───────── Decompositions.h ─────────
class SquareMatrixDecomposition_ : noncopyable
{
    virtual void XMultiplyLeft_af
        (const Vector_<>& x, Vector_<>* b)
    const = 0;
    virtual void XMultiplyRight_af
        (const Vector_<>& x, Vector_<>* b)
    const = 0;
    virtual void XSolveLeft_af
        (const Vector_<>& b, Vector_<>* x)
    const = 0;
    virtual void XSolveRight_af
        (const Vector_<>& b, Vector_<>* x)
```

---

[6] Private inheritance does not work for this purpose, because C++ will find the inheritance – which is inaccessible – before checking for a conversion operator.

```
        const = 0;
15  public:
        virtual ~SquareMatrixDecomposition_() {}
        virtual int Size() const = 0;     // of the matrix
        // these nonvirtual methods handle aliasing:
        void MultiplyLeft(const Vector_<>& x, Vector_<>* b) const;
20      void MultiplyRight(const Vector_<>& x, Vector_<>* b) const;
        void SolveLeft(const Vector_<>& b, Vector_<>* x) const;
        void SolveRight(const Vector_<>& b, Vector_<>* x) const;
    };
```

The `MultiplyLeft` and `MultiplyRight` functions compute Ax and xA, respectively, where A is the *original* matrix; `SolveLeft` and `SolveRight` solve $Ax = b$ and $xA = b$. All these are implemented as virtual private expert alias-free functions (as seen in their names), and then accessed through a public interface which handles any aliasing. Behind this interface can live LU, QR and SVD decompositions from *Numerical Recipes*, BLAS, or the proprietary vendor of your choice.

A particular decomposition, necessary for efficient implementation of Markov-chain models, is eigenvalue decomposition of a non-symmetric matrix to permit exponentiation of a matrix. Since not all decompositions can support this, we must extend the interface, using a mixin:

```
———————————————— Decompositions.h ————————
    class ExponentiatesMatrix_
    {
    public:
        virtual void ExpAT
5           (double t, SquareMatrix_<>* dst)
            const = 0;
    };
```

Thus `Eigensystem_` will derive from `SquareMatrixDecomposition_` and also from `ExponentiatesMatrix_`; a function requiring exponentiation would use `dynamic_cast` to access the extra functionality.

4.6 Decompositions (Symmetric)

The most widely used decompositions, mainly because of their connection to covariance matrices, are those of symmetric matrices. The interface is unsurprising:

```
 ──────────────────── Decompositions.h ────────────────────
   class SymmetricMatrixDecomposition_
          : public SquareMatrixDecomposition_
   {
       virtual void XMultiply_af
 5        (const Vector_<>& x,
          Vector_<>* b)
       const = 0;

       void XMultiplyLeft_af
10        (const Vector_<>& x, Vector_<>* b) const override
       {   XMultiply_af(x, b);   }
       void XMultiplyRight_af
          (const Vector_<>& x, Vector_<>* b) const override
       { XMultiply_af(x, b); }
15
       virtual void XSolve_af
          (const Vector_<>& b, Vector_<>* x)
       const = 0;
       void XSolveLeft_af
20        (const Vector_<>& b, Vector_<>* x) const override
       { XSolve_af(b, x); }
       void XSolveRight_af
          (const Vector_<>& b, Vector_<>* x) const override
       { XSolve_af(b, x); }
25
   public:
       virtual int Rank() const { return Size(); }
       virtual Vector_<>::const_iterator MakeCorrelated
          (Vector_<>::const_iterator iid_begin,
30        Vector_<>* correlated)
       const = 0;
       // These nonvirtual methods handle aliasing:
       void Multiply(const Vector_<>& x, Vector_<>*b) const;
       void Solve(const Vector_<>& b, Vector_<>* x) const;
35 };
```

The MakeCorrelated function consumes i.i.d normal deviates from an input vector and populates a vector of correlated multivariate normal deviates whose covariance matrix is the source of the decomposition. It returns a pointer to the first unused i.i.d. deviate after the ones it used, which simplifies chaining of such correlators.

Cholesky and eigenvalue decompositions are the most important instances of this type. While the Cholesky decomposition is faster, both to construct and in MakeCorrelated, we will often prefer the eigenvalue decomposition. The latter provides a somewhat meaningful ordering of the input i.i.d. normal deviates in order of importance, which aids variance reduction in Monte Carlo using quasi-random sequences (see Sec. 7.8.5). Often the eigenvalue decomposition can be truncated after a few modes, greatly increasing the speed of simulation with minimal loss of accuracy (naturally, a truncated decomposition can no longer `Solve()` anything).

The important advantage of the Cholesky decomposition is not its speed, but its stability. In cases where truncation is not feasible and the eigenvalue is not a measure of importance – such as the joint evolution of interest rates, FX and equities in a hybrid model – we will prefer the Cholesky decomposition.

4.7 Decompositions (Sparse)

A large matrix with relatively few nonzero elements is called *sparse*, and there is a substantial literature of mathematical methods for such matrices. The sparse matrices most useful to us are square as well. A particular task of these matrices is to compute the *Q-form* $JW^{-1}J^T$ used in underdetermined search; see Sec. 7.4. We define them in **namespace Sparse**:

```
———————————— Sparse.h ————————————
class SymmetricDecomposition_
      : public SymmetricMatrixDecomposition_
{
public:
5     // form J^T A^{-1} J for given J
      virtual void QForm
          (const Matrix_<>& J,   SquareMatrix_<>* dst)
      const;
};
10
class Square_ : noncopyable
{
public:
      virtual int Size() const = 0;
15
      virtual void MultiplyLeft
          (const Vector_<>& x, Vector_<>* b) const = 0;
      virtual void MultiplyRight
          (const Vector_<>& x, Vector_<>* b) const = 0;
20
      virtual bool IsSymmetric() const = 0;
      // returns a SymmetricDecomposition_ if possible
      virtual SquareMatrixDecomposition_* Decompose()
      const = 0;
25
      // element access
      virtual const double& operator()(int row, int col)
      const = 0;
```

```
     virtual void Set(int row, int col, double val) = 0;
30   virtual void Add(int row, int col, double val)
     {
        Set(row, col, val + operator()(row, col));
     }
};
```

The default implementation of QForm relies on repeated calls to SolveLeft, which is inherited from Sparse::Decomposition_. For some decompositions, such as banded Cholesky, we can override this implementation to reduce the number of calls to SolveLeft by a factor of two.

A sparse matrix itself does not allow access to the details of its layout; we manipulate it through the Set and Add functions. By default, Add is implemented using Set, but we make it virtual to permit optimization. We prefer to return a const double & from operator(), rather than a double, so that incorrect code like A(i,j)=a will not compile.

4.7.1 Tridiagonal Matrices

The most familiar sparse matrices are tridiagonal, where only elements immediately adjacent to the diagonal are nonzero. A tridiagonal matrix is thus represented by three vectors. In our implementation, rather than have dummy entries in the vectors representing the off-diagonal elements, we allocate only $n - 1$ elements for them.

```
───────────────────────── Banded.cpp ─────────────────────────
class Tridiagonal_ : public Sparse::Square_
{
   Vector_<> diag_, above_, below_;
public:
   Tridiagonal_(int size);
5   int Size() const { return diag_.size(); }
   bool IsSymmetric() const { return above_ == below_; }
   // ...
```

The accessor At returns the address of an element in one of the member vectors, or nullptr if the element is too far off-diagonal. In the latter case the Set and Add functions will throw if the input val is nonzero, while operator() will return a reference to a local immutable zero value.

This implementation is completely local to Tridiagonal_'s source file; in fact, so is Tridiagonal_ itself. Only a factory function, with a return value of type Sparse::Square_*, appears in the header. Decompose sends the matrix data to an object of a new type:

```
─────────────────────── Banded.cpp ───────────────────────
struct TriDecomp_ : Sparse::Decomposition_
{
    Vector_<> diag_, above_, below_;
    Vector_<> betaInv_;

    TriDecomp_(diag_, above_, below_)
    : betaInv_(TridagBetaInverse(diag_, above_, below_))
    { }

    int Size() const {return diag_.size();}
    void XMultiplyLeft_af(const Vector_<>& x, Vector_<>* b)
        const
    {
        assert(x.size() == Size());
        TriMultiply(x, diag_, above_, below_, b);
    }
    void XMultiplyRight_af(const Vector_<>& x, Vector_<>* b)
    // ...
};
```

The local free function `TriMultiply` performs either left- or right-multiplication, depending on the order of input arguments. A similar function `TriSolve` implements the decomposition and backsubstitution loop, which may be familiar from *Numerical Recipes*. The difference is that we can take `beta` as an input, having formed it in the constructor; as a result, no temporary vector is needed. Thus the up-front effort of allocation pre-computing β^{-1} pays for itself in a single use.

If the computation of β^{-1} fails due to the lack of pivoting, `Tridiagonal_::Decompose` could revert to some more robust method (returning a different subclass of `Sparse::Decomposition_`). We do not implement this refinement.

We can implement a simplified variant of `TriDecomp_` for the symmetric case, and return it from `Tridiagonal_::Decompose` when the original matrix is symmetric. This is useful if the calling code requires a symmetric decomposition (though the caller must `dynamic_cast` the return value to the desired type).

4.7.2   *Band Diagonal Matrices*

Band-diagonal matrices generalize tridiagonals, allowing nonzero entries a (small) fixed distance above or below the diagonal. Our class definition and constructor are, as usual, concealed behind a factory function.

```
────────────────────── Banded.h ──────────────────────
namespace Sparse
{
    Square_* NewBandDiagonal
        (int size, int n_above, int n_below);
}
```

Our implementation closely follows that of `Tridiagonal_`, but the `Decompose` function must actually perform a banded LU decomposition.[7]

There is one important optimization. If the matrix being decomposed is symmetric, then we can attempt a Cholesky decomposition; this, if it succeeds, gives a more compact representation (taking $\frac{1}{3}$ as much memory) and also supports a more efficient implementation of `QForm`. For, if $A = LL^T$, then $J^T A^{-1} J = (L^{-1}J)^T(L^{-1}J)$; thus we proceed by computing $L^{-1}J$ immediately.

```
───────────────────── Banded.cpp ─────────────────────
void BandedCholesky_::QForm(const Matrix_<>& j,
    SquareMatrix_<>* form) const
{
    assert(j.Cols() == Size());
    Vector_<Vector_<>> tm(j.Rows());
    // compute L^{-1} J
    for (int ii = 0; ii < j.Rows(); ++ii)
    {
        tm[ii].Resize(Size());
        Copy(j[ii], &tm[ii]);
        BandedLSolve(vals_, tm[ii], &tm[ii]); // in-place
    }
    // compute result
    form->Resize(j.Rows());
    for (int io = 0; io < j.Rows(); ++io)
        for (int k = 0; k <= io; ++k)
            (*form)(io, k) = (*form)(k, io)
                = InnerProduct(tm[io], tm[k]);
}
```

Our use of row-major storage for matrix elements means that iteration along rows is substantially more efficient than along columns. The `vals_` of a `BandedCholesky_` decomposition are oriented so that the inner loop of `BandedLSolve` requires no column iteration; thus it can be made more efficient than the corresponding `BandedLTransposeSolve`. The result is that this version of `QForm`, which requires no computations using L^T, is well over twice as efficient as the default.[8]

---

[7]For which we will likely look to *Numerical Recipes*.
[8]This is an example of the principles laid out in Sec. 2.3.

4.7.3 *SLAP Format*

General sparse matrices, whose nonzero elements are not guaranteed to lie near the diagonal, require more sophisticated memory management. The most efficient format for our purposes is a `Vector_<>` of diagonal elements, plus another vector containing for each row (thus our storage is called *row-indexed*) an array of nonzero values and their column locations. This latter is more efficiently implemented using a `Vector_` than a `list`:

```
                            SLAP.cpp
class SlapMatrix_ : public Sparse::Square_
{
    Vector_<> diag_;
    Vector_<Vector_<pair<int, double>>> offDiag_;
5
    void XMultiplyLeft_af
        (const Vector_<>& x, Vector_<>* b) const override;
    // ...
    const double& operator()
10      (int i_row, int i_col) const override;
    // ...
};
```

Our algorithms are based on those of the *Sparse Linear Algebra Package*, or *SLAP*.

Methods like `SolveLeft` can no longer be implemented exactly; we must rely on the iterative *conjugate gradient* and, for nonsymmetric matrices, *biconjugate gradient* methods. These are made more efficient by *preconditioning*: rather than solve $Ax = b$ iteratively, we choose a matrix $\tilde{A}$ for which $\tilde{A}\tilde{x} = b$ can be solved efficiently, then solve $\tilde{A}^{-1}Ax = \tilde{A}^{-1}b = \tilde{x}$ iteratively using conjugate gradient.

We provide a mixin which will allow matrices to supply a preconditioner:

```
                            BCG.h
class HasPreconditioner_
{
public:
    virtual void PreconditionerSolveLeft
5      (const Vector_<>& b, Vector_<>* x) const = 0;
    virtual void PreconditionerSolveRight
        (const Vector_<>& b, Vector_<>* x) const = 0;
};
```

The biconjugate gradient method, which can be found in *Numerical Recipes*, implements `SolveLeft` in terms of `MultiplyLeft`. We implement it as a free function

```
──────────────── BCG.cpp ────────────────
void BCGSolveLeft
   (const Sparse::Square_& A,
    const Vector_<>& b,
    double tolerance,
 5  int max_iterations,
    Vector_<>* x);
```

We check for a preconditioner using `dynamic_cast`; thus a run-time decision for a class instance not to use preconditioning (if at compile time we have supplied the inheritance from `HasPreconditioner_`) must be implemented by supplying `Copy(b, x)` or `*x=b` in the body of the `PreconditionerSolve` functions.

A useful helper is the ephemeral struct:
```
──────────────── BCG.cpp ────────────────
struct XPrecondition_
{
   const HasPreconditioner_* a_;    // we do not own this
   XPrecondition_(const Sparse::Square_& a)
 5    : a_(dynamic_cast<const HasPreconditioner_*>(&a)) {}
   void Left(const Vector_<>& b, Vector_<>* x) const
   {
      if (a_)
         a_->PreconditionerSolveLeft(b, x);
10    else if (x != &b)
         Copy(b, x);
   }
   // ... similarly for Right
};
```

This lets the solver treat sparse matrices, with and without preconditioning, uniformly.

The simplest way to support `SolveRight` is to call `SolveLeft` with a transposed matrix. Since we do not happen to need transposed matrices in general, we use an ephemeral local struct which holds a reference to `A`:
```
──────────────── BCG.cpp ────────────────
struct XSparseTransposed_
      : public Sparse::Square_, public HasPreconditioner_
{
   const Sparse::Square_& a_;
 5  XPrecondition_ p_;
   XSparseTransposed_(const Sparse::Square_& a)
         : a_(a), p_(a) {}

   int Size() const {return a_.Size();}
10  void XMultiplyLeft_af(const Vector_<>& x,
```

```
        Vector_<>* b) const { a_.MultiplyRight(x, b); }
    void XSolveLeft_af(const Vector_<>& b,
        Vector_<>* x) const { UNREACHABLE; }
    void PreconditionerSolveLeft
        (const Vector_<>& x, Vector_<>* b) const
    { p_.Right(x, b); }
    // ... Right functions are all UNREACHABLE
};
```

Now we can write BCGSolveRight, which just calls BCGSolveLeft sending XSparseTransposed_(a) as the matrix. Since BCGSolveLeft does not itself call any Solve function, XSparseTransposed_ should not implement them.

We provide a template implementation to support both left- and right-multiplication. The implementation gives an idea of how the SLAP layout is used:

——————————— *SLAP.cpp* ———————————
```
template<bool transpose> void XMultiply
    (const Vector_<>& x,
    Vector_<>* b)
const
{
    const int n = Size();
    b->Resize(n);
    Transform(x, diag_, multiplies<double>(), b);
    // now add off-diagonal
    for (int ii = 0; ii < n; ++ii)
        for (auto l_v : offDiag_[ii])
            (*b)[transpose ? l_v.first : ii] += l_v.second *
                x[transpose ? ii : l_v.first];
}
```

When accessing an element (*e.g.*, within Set), we must distinguish diagonal from off-diagonal entries, and also handle both modification of an existing element and addition of a new one. As for the tridiagonal matrix, we support the const and non-const cases with a single templated free function, to which the SLAP matrix will pass its data. We use C++11's trailing return type to avoid making the return type a template parameter (since it cannot be deduced at the function call point):

——————————— *SLAP.cpp* ———————————
```
template<class D_, class O_> auto SLAPElement
    (D_& diag, O_& off_diag, int i_row, int i_col)
-> decltype(&diag[0])
{
    if (i_row == i_col)
        return &diag[i_row];
```

```
     auto row = off_diag[i_row];
     for (auto l_v : row)
        if (l_v.first == i_col)
10          return &l_v.second;
     return nullptr;
  }
```

Here we do not assume the off-diagonal elements are sorted in column order. Doing so would let us use binary rather than linear search here, slightly accelerating queries of the SLAP matrix at the expense of some complexity. In practice, we expect the number of off-diagonal elements per row to remain small, and the computation time to be dominated by XMultiply (which is called iteratively by the conjugate gradient solver), so we do not implement this optimization.

This function supports operator() directly, and support Set and Add except when a new element must be added. Since off-diagonal elements are unordered, they can simply be pushed as needed:

SLAP.cpp
```
void Set(int i_row, int i_col, double val) override
{
    if (double* dst = SLAPElement(diag_, offDiag_, i_row, i_col))
        *dst = val;
5   else
        offDiag_[i_row].push_back(make_pair(i_col, val));
}
```

Note that a call to Set or Add can invalidate the returned reference from operator(); this is not a problem in practice.

4.7.4 The Symmetric Case

Oddly, *Numerical Recipes* provides code for biconjugate gradient but not the symmetric specialization to conjugate gradient. The latter routine can dispense with half of the preconditioner steps and some vector arithmetic, and is worthwhile to implement as an optimization. The parameter x on input contains a guess at the solution; on output it contains the true solution.

BCG.cpp
```
void Sparse::CGSolve
    (const Sparse::Square_& A,
    const Vector_<>& b,
    double tolerance,
5   int max_iterations,
    Vector_<>* x)
```

```
{
    const int n = A.Size();
    assert(b.size() == n && x->size() == n);
10  assert(!IsZero(tolerance) && max_iterations > 0);

    double tNorm = tolerance * sqrt(InnerProduct(b, b));
    XPrecondition_ precondition(A);
    Vector_<> r(n), z(n), p(n);
15  A.MultiplyLeft(*x, &r);
    Transform(b, r, minus<double>(), &r);   // r = b - Ax
    double bkden;
    for (int ii = 0; ii < max_iterations; ++ii)
    {
20      precondition.Left(r, &z);
        const double bknum = InnerProduct(z, r);
        p *= ii > 0 ? bknum / bkden : 0.0;
        p += z;
        bkden = bknum;
25      A.MultiplyLeft(p, &z);
        const double ak = bknum / InnerProduct(z, p);
        Transform(x, p, LinearIncrement(ak));
        Transform(&r, z, LinearIncrement(-ak));
        if (sqrt(InnerProduct(r, r)) <= tNorm)
30          return;
    }
    THROW("Exhausted iterations in CGSolve");
}
```

The comparison of this routine with the original is also a good demonstration of the expressive power of higher-level programming.

One useful preconditioner for symmetric sparse matrices is the *incomplete Cholesky decomposition*. Here we attempt a Cholesky decomposition, with the constraint that the decomposition may have nonzero values only where the source matrix itself has a nonzero value. This decomposition is, as its name suggests, incomplete; $LL^T \neq A$. An alternative preconditioner is the tridiagonal part of A, ignoring all other nonzero entries; this is less similar to A but easy to form and rapid in application. These three possibilities (incomplete Cholesky, tridiagonal part, or no preconditioner) must be tested in realistic cases to see which provides the best performance; there is no clear winner in all scenarios.

Band diagonal matrices should be favored when possible due to the speed and precision of the banded LU and Cholesky decompositions. The generic SLAP form is used only in some advanced applications, such as simultaneous underdetermined calibration of multiple coupled surfaces.

4.8 Decompositions (Other)

A diagonal or lower-triangular matrix is already "decomposed" enough for our purposes; we can easily `Multiply` and `Solve` with it. It is useful to provide utility functions which implement `SquareMatrixDecomposition_` for such matrices, which often arise in Jacobians for chaining of risk sensitivities. The header file does not need to display the concrete class, just a factory function:

```
_____ DecompositionsMisc.h _____
SymmetricMatrixDecomposition_* DiagonalAsDecomposition
    (const Vector_<>& diag);
SquareMatrixDecomposition_* LowerTriangularAsDecomposition
    (const SquareMatrix_<>& src);
```

Singular value decomposition, which supports linear fitting, is another useful tool. We do not display it here, but there is one failing of the common implementation which we should point out. The function

```
xInv = x > xMin ? 1.0 / x : 0.0;
```

which is often used in place of the simple $1/x$ when inversion of the singular values is called for, is gratuitously discontinuous. Continuity is necessary for small input changes to lead reliably to small value changes, and desirable in many other ways. Thus we use instead

```
xInv = x > xMin ? 1.0 / x : x / Square(xMin);
```

or even

```
xInv = x / (Square(x) + Square(xMin));
```

where xMin is, as usual, a user-supplied fraction of the maximal singular value.

This page intentionally left blank

Chapter 5

Persistence and Memory

To support both debugging of complex tasks and distributed computing, we will need to provide object persistence – the ability to save an object or transfer it to a different process, generally by writing it to a file or database.

Some languages offer built-in support for this, or have a sufficiently simple object model that a general-purpose persistence library can be built. In the Python language, for example, any object is essentially a nested hash table, and the "pickle" module provides object-to-file conversions for most objects. Erlang goes even further, with built-in concurrency and little distinction between persisted and in-process data. However, partly because of the constraints required to support this universality, these languages are not really suitable for our purposes.

More general-purpose languages often come with libraries which support persistence of objects built within that library: examples include CL-STORE for LISP and MFC for C++.

However, there are significant advantages to writing our own persistence methods, which as far as I am aware have not been replicated in any widely available library. Foremost among these is the automatic maintenance of documentation, just as for our generated public interface. Also, serialization is nearly conceptually identical to inspection, and we will support both without duplicative code.

5.1 Storage

We will define a *storable* base class for objects which can be saved and recovered ("persisted" in the regrettable lexicon of quants). Writing functions which save to some concrete storage mechanism – *e.g.*, XML – is a common mistake. Obviously, this practice binds the code to a single type of

71

storage and wastes the opportunity to create new implementations behind an abstract interface. In practice, it also leads to the proliferation of different persistence idioms; these might all be valid, but they act in concert to destroy the uniformity of stored data and of persistence code.

Thus `Storable_` should know only about an abstraction of archiving:

```
───────────────── Storable.h ─────────────────
    namespace Archive { class Store_; }

    class Storable_ : noncopyable
    {
5   public:
        const String_ type_, name_;
        virtual ~Storable_();
        Storable_(const char* type, name_) : type_(type) {}
        // support actual storage of these things
10      virtual void Write(Archive::Store_& dst) const = 0;
    };
```

The `type` is supplied as a C-style string, rather than a `String_`, simply to check that the two inputs to the constructor are not inadvertently swapped.

`Archive::Store_` is the interface through which an object writes itself. Classes derived from `Storable_` must use this interface to implement `Write`, but files which *use* storable objects do not need it; thus we place it in a separate file.

```
───────────────── Archive.h ─────────────────
    class Store_ : noncopyable
    {
    public:
        virtual ~Store_();
5       virtual void SetType(const String_& type) = 0;
        virtual bool StoreRef(const Storable_* object) = 0;
        // true -> stored; false -> created new tag, awaiting store

        virtual Store_& Child(const String_& name) = 0;
10      Store_& Element(int index); // implemented through Child()

        virtual void operator=(double val) = 0;
        virtual void operator=(const String_& val) = 0;
        virtual void operator=(const Vector_<>& val) = 0;
15      virtual void operator=(const Vector_<int>& val) = 0;
        virtual void operator=(const Vector_<bool>& val) = 0;
        // ...
        virtual void operator=(const Matrix_<Cell_>& val) = 0;
        virtual void operator=(const Dictionary_& val) = 0;
20  };
```

The operator= functions set the archive to contain one concrete datum; such an archive needs no further type information, and should never be given any.

The function StoreRef is needed to support storage by reference, so that shared sub-objects can be written only once and afterwards a unique identifying tag can be substituted.[1] It returns true if such a reference can be found; then the write is complete. More likely it will return false, meaning that the object is not written, and we *must* now write it. This demand – write if and only if StoreRef returns false – simplifies the writer's implementation, because StoreRef can do any necessary writing when it succeeds, and can create a tag for future reference when it fails.

Rather than rely on developers to infallibly honor this restriction, we provide a utility function:

```
────────────────────── Storable.cpp ──────────────────────
void Archive::Utils::SetStorable
   (Archive::Store_& dst,
    const String_& name,
    const Storable_& value)
{
    auto& child = dst.Child(name);
    if (!child.StoreRef(&value))
      value.Write(child);
}
```

This is the sole caller of StoreRef, so we can give it friend status and make StoreRef a private function.[2] A partial specialization of Archive::Utils::Set forwards all Handle_ values to SetStorable.

This gives us the tools to write to some abstract archive any of the fundamental types in the interface above, or any object composed of members which can themselves be archived. One subtlety is that the type_ of a Storable_ object will generally not be the type sent to Store_::SetType. The former is mainly a hint or comment so users can quickly know the purpose of an object; the latter is specific to a particular concrete class (and, in a mature system, to a version of the object's serialization; see Sec. 5.3).

---

[1] The tag is seen only by the archive, not by the object.
[2] This same function can be made the sole caller of Storable_::Write, so we can make the latter private as well. We choose not to implement this because it entails a "long-distance friendship" across component files.

5.2 Extraction

We can reconstitute storable instances from an archive, through a query
interface which is the mirror image of the writing interface above (also in
namespace Archive).

```
_____ Archive.h _____
class View_ : noncopyable
{
public:
    virtual ~View_();

    // query fundamental types
    virtual double AsDouble() const = 0;
    virtual int AsInt() const = 0;
    virtual String_ AsString() const = 0;
    virtual Dictionary_ AsDictionary() const = 0;
    virtual Vector_<> AsDoubleVector() const = 0;
    virtual Vector_<int> AsIntVector() const = 0;
    virtual Vector_<bool> AsBoolVector() const = 0;
    // ...
    virtual Matrix_<Cell_> AsCellMatrix() const = 0;

    // query composite types
    virtual String_ Type() const = 0;    // empty for atoms
    virtual const View_& Child(const String_& name) const = 0;
    virtual bool HasChild(const String_& name) const = 0;
    const View_& Element(int index) const;
    bool HasElement(int index) const;
    // notify of unexpected children
    virtual void Unexpected(const String_& child_name) const = 0;
    // check for ready-built object (returns reference in 'built')
    virtual Handle_<Storable_>& Known(Built_& built) const = 0;
};
```

The extraction of fundamental types is straightforward: the only guar-
antee we need is that an archive leaf element created with a particular
operator= in Archive::Store_ must return the same value from the corresponding query function. In practice other queries will likely succeed as
well: for example, a number will be stored in an JSON file as text, and thus
could be read out as a String_ unless we clutter the file with extra type
information. It is better to keep the archive simpler, since the "accidental"
success of a query does no practical harm.

This list of fundamental types is not exhaustive; for instance, there
is no AsDateMatrix. It may come to pass that some object requires
the serialization of a Matrix_<Date_>; at that time we would have to add
the necessary functionality to all the Archive classes.

A composite object must have a unique `Type` which identifies the recipe for reconstituting it from an archive. These recipes will be placed into a global table of "builders" which are found and invoked by a master extractor:

```
───────────────────────── Archive.cpp ─────────────────────────
Handle_<Storable_> Archive::Extract
    (const Archive::View_& src,
     Built_& built)
{
    Handle_<Storable_>& retval = src.Known(built);
    if (!retval)
    {
        const String_& type = src.Type();
        REQUIRE(!type.empty(), "No type in store");
        NOTICE(type);
        auto pb = TheBuilders().equal_range(type);
        REQUIRE(pb.first != pb.second, "No builder found");
        REQUIRE(pb.first == --pb.second, "Builder is not unique");
        retval.reset(pb.first->second->Build(src, built));
    }
    return retval;
}
```

The first part of this function relies on `View_::Known`; this is the complement to `Store_::StoreRef`, checking whether the `View_` contains a reference which can be used to locate an already-built object.[3] During extraction of a nested object, we accumulate such objects in the `Built_` structure, which is just a map from `String_` keys to `Storable_ Handle_s`.

The bulk of `Extract` handles the usual case, where the object is not `Known` and must be constructed. For this purpose we introduce a new base class, `Archive::Reader_`, which intermediates between the `View_` and the final object; the reader's role is to marshal data from the `View_` and then use it to `Build` the final `Storable_` object. Thus we arrive at the following lifecycle for storable objects:

- An original storable object writes itself, using the abstract `Store_` interface.
- The concrete implementation of `Store_` captures the resulting information, and places it in some persistent storage.

---

[3]This relies on the data's being read from the archive in the same order as that in which it was written. In practice, a single-threaded reading and writing mechanism will satisfy this constraint without effort on our part, but we have closed the door to multi-threaded reading of a single object.

- A `View_` is a reading interface for exactly the same persistent data.
- The `View_` discloses a `Type` which is used to find a corresponding `Reader_`.
- The `Reader_` queries the `View_` for the data it requires, then uses that to `Build` the final object.

At the end of this process, the final object should be identical to the original.[4] This requires consistency between each derived `Store_` and its corresponding `View_`, and between the read/build process and the original object's `Write`. We will enforce the latter, and make the developer's life much easier, by machine-generating reading and writing code from a simple mark-up description of the contents; see Sec. 5.3. To support objects storing overcomplete or otherwise preprocessed data (*e.g.*, the Vasicek-Hull-White `Vol_` in Sec. 13.1), we allow the option to suppress the machine generation of `Build` and instead supply a handwritten version. (Note that the machine-generated version requires the members to be presented in mark-up in the same order as that supplied to the `Storable_` object's constructor.)

In principle we could construct a `Storable_` directly from the archive; but this would require all `Storable_` objects to implement a suitable constructor interface with a specific argument ordering. It would also constrain the internal layout of each `Storable_`, which would have to hold all its members in the canonical way expected by the archive. The role of `Reader_` is to hold the necessary data and simultaneously obey all these constraints; by having the `Reader_` double as a builder, we also gain a convenient way to maintain a registry of builders.

5.2.1 *Example: Linear Interpolant*

To see the write-read cycle in action, we look at a simple linear interpolator. The only data members of such an interpolator are the vectors `x` and `f` containing the independent and dependent variables, respectively; and the `name` attribute of any `Storable_` object. Such a simple class hardly requires versioning, but we supply it here for completeness. The class's contribution to its serialization is a `Write` function, with a body like:

```
     dst.SetType("Interp1Linear_v1");
   if (name != String_())
      dst.Child("name") = name_;
   dst.Child("x") = x_;
5  dst.Child("f") = f_;
```

---

[4]Or, at worst, might have some empty values replaced with defaults.

This sets a concrete type string for the archive, and creates child archives containing the members as necessary. By introducing some simple helper functions in **namespace Archive::Utils**, we can rewrite this as:

```
————————————— MG_Interp1Linear_Write.inc —————————————
    using namespace Archive::Utils;
    dst.SetType("Interp1Linear_v1");
    SetOptional(dst, "name", name);
    Set(dst, "x", x);
5   Set(dst, "f", f);
```

Our aim here is not the slight decrease in verbosity, but to give the writing code a very regular structure, making it easy to machine-generate.

Next we require a **Reader_** which can extract exactly this data from an existing archive. To avoid name clashes, we enclose it within **namespace Interp1Linear_v1**.

The unique capability of the reader, of course, is the extraction of data sufficient to build an **Interp1Linear_**. We provide this as an additional constructor, which extracts data from the archive, and a separate **Build** function:

```
————————————— MG_Interp1Linear_Read.inc —————————————
    Reader_(const Archive::View_& src, Archive::Built_& share)
    {
        using namespace Archive::Utils;
        NOTE("Reading Interp1Linear_v1 from store");
5       assert(src.Type() == "Interp1Linear_v1");
        GetOptional(src, "name", &name_,
            mem_fun_ref(&Archive::View_::AsString));
        Get(src, "x", &x_, mem_fun_ref(&View_::AsDoubleVector));
        Get(src, "f", &f_, mem_fun_ref(&View_::AsDoubleVector));
10  }
    Interp1Linear_* Build() const
    {
        return new Interp1Linear_(name_, x_, f_);
    }
```

An overloaded version of **Build** combines these tasks to produce the object we want:

```
————————————— MG_Interp1Linear_Read.inc —————————————
    Interp1Linear_* Build(const Archive::View_& src,
        Archive::Built_& share) const
    {
        return Reader_(src, share).Build();
5   }
```

5.2.2 *Reader Registry*

Now that we can extract `Data_` of one particular type, we must prepare for objects whose exact type is unknown. But these concrete `Data_` types are scattered across the code, and isolated in individual objects. A generic reader must be able to find them, which of course is the purpose of the `Type` tag of an archived object; and to invoke the function which reads and translates archived data.

Each reader provides such a function, embedded in its vtable; thus the global registry of builders must hold reader instances which will provide the necessary vtable access. We accomplish this by giving each reader a constructor which registers the constructed instance:

```
 ─────────────── MG_Interp1Linear_Read.inc ───────────
     Reader_(void (*register_func)(const String_&,
         const Archive::Reader_*))
     {
         register_func("Interp1Linear_v1", this);
5    }
// ...
static Reader_ TheData(Archive::Register);
```

By using these constructors to create a static instance of each `Reader_`, we ensure their presence in the builder registry.[5]

We provide a central singleton registry (a `map<String_, reader_t>`), and export the function `Register` as shown above; it will be called for each reader at DLL load time.

Now, given an archived object, a centralized function can look at its `Type` and dispatch to the appropriate reader; this is exactly how `Archive::Extract` operates in Sec. 5.2 above.

5.3   Code Generation

The code displayed in Sec. 5.2.1, particularly the archive reader, is verbose and error-prone to write. This is because it is a low-level representation of high-level concepts. We can represent the same information in a clear and compact form:

```
 ──────────────── Interp1.h ────────────
/*IF----------------------------------------------------------
storable Interp1Linear
    Linear interpolator on known values in one dimension
```

---

[5]The builder registry will thus be invalidated during DLL unloading, as these objects are destroyed; I am unable to construct a use case where this is a problem.

```
  version 1
5 &members
  name is ?string
  x is number[]
  f is number[]
  -IF-----------------------------------------------------*/
```

If we later decide to use a different object layout for the same `Storable_` type, we will leave this mark-up (and the associated code) and add an additional mark-up block with a different `version`.

This provides enough information to generate all our example code. Within `namespace Interp1Linear_v1`, we provide a free function `XWrite` which performs the write as described above; a handwritten `Write` function in `class Interp1Linear_` forwards the member data to `XWrite`.

The reading and building functions given above are all collected into a `Reader_` class in this same namespace. If the mark-up contains the keyword `manual`, then the `Build()` function is not provided and must be handwritten.

At this point, persistence requires only one or two pieces of handwritten C++ code: the forwarding function `Write`, and sometimes a `Build` method.

The machine-generated code uses `Archive::Utils::Set`, or the template helpers `SetOptional` and `SetMultiple`, for every member datum regardless of its type. For each fundamental type, `Set` forwards to the corresponding member function in `Store_`; for `Storable_` objects (and `Handle_s` thereto), we provide the template specialization shown in Sec. 5.1.

Note that we do not distinguish among various subtypes of `Storable_` in the reading or extraction process. But when we deal with nested objects, generally the member (inner) object's type will be specified; *e.g.*, an interest rate model wants to contain a `YieldCurve_`, not just a generic `Storable_`.

We address this need by providing a templated extractor function to read member objects:

```
_____ Archive.h _____
  template<class T_ = Storable_> struct Builder_
  {
      Built_& share_;
      const char* name_;
5     const char* type_;
      Builder_(Built_& s, name_, type_) : share_(s) {}
      Handle_<T_> operator()(const View_& src) const
      {
          NOTICE2("Child name", name_);
10        Handle_<Storable_> object = Extract(src, share_);
          // add type
```

```
         NOTICE2("Expected type", type_);
           Handle_<T_> retval = handle_cast<T_>(object);
           REQUIRE(retval, "Object has wrong type");
15           return retval;
       }
   };
```

Now the code to extract and type-check a member object looks like:

```
Get(src, "yc", &yc_, Archive::Builder_<YieldCurve_>
      (share, "yc", "YieldCurve"));
```

The extra mention of `"yc"` and `"YieldCurve"` is not needed for functionality, but allows us to construct better error messages. Since the `name_` will stay on the exception-message stack throughout the call to `Extract` (see Sec. 3.6), a full record of the path through the object will be attached to any error message.

Note that the machine-generated code is centered around the machine-generated `Reader_` type, and our class interacts with it through `Write` and `Build` functions which we may choose to hand-write. This preserves implementation freedom for the working classes.

5.4   A Display Interface

A particular archive of interest is one which converts (*"splats"*) a `Storable_` to a two-dimensional tableau, which can then be displayed in a spreadsheet or stored as a tab-separated file. For this we need concrete implementations of `Archive::Store_` and `Archive::View_`; we will (almost) call them `Splat_` and `Unsplat_` respectively.

5.4.1   *Storage*

In fact, our implementation relies on sharing ephemeral references, so the writer's type is `XSplat_`. This will, once fully populated, convert itself to a two-dimensional table. The table's size will be that of what it stores (if it contains a concrete datum), or the combined sizes of its children or elements.

Consider the virtual members of `Archive::Store_` (in Sec. 5.1). The overloaded versions of `operator=` will set internal data to contain the right-hand side of the assignment. All these concrete types can be mapped onto a two-dimensional `Matrix_<Cell_>`, so this is the type we will use for storage. (A `Dictionary_` can be represented as a two-column array of keys and values, or as a comma- or semicolon-separated string.)

Alternatively, the archive might have to contain a `Storable_` composite object. In this case, it will receive a type and some set of `Child` or `Element` data, for each of which it will create a new `XSplat_`.

Finally, the function `StoreRef` depends on information shared across the entire archive. We can implement this as a member `shared_ptr`, created by the parent archive and passed to children, or as data outside the class to which each `XSplat_` object receives a reference. The latter approach is clearer and more concise, but we must ensure that the reference is valid as long as it is needed, so we can never let a `XSplat_` escape the reference's scope. Thus the implementation is locally namespaced within a source file:

```
———————————————— Splat.cpp ————————————————
   struct XSplat_ : Archive::Store_
   {
       String_ tag_;
       String_ type_; // for storable reader/builder
5      // String_ dataType_;
       map<String_, shared_ptr<XSplat_>> children_;
       map<const Storable_*, String_>& sharedTags_;
       Matrix_<Cell_> val_;
```

A concrete value is stored into our own `data_` table, while composite values declare a type and then are stored as `children_`. Elements are just children with numeric names.

The `dataType_` string could be used to add type information, e.g. to invalidate the operation of writing an integer to an archive and then viewing it as a double. We do not fear this failure mode, which is both unlikely and benign, so we do not implement this layer of type checking.

```
———————————————— Splat.cpp ————————————————
       XSplat_& Child(const String_& name) override
       {
           shared_ptr<XSplat_>& retval = children_[name];
           if (!retval.get())
5              retval.reset(new XSplat_(sharedTags_));
           return *retval;
       }
```

Assignment to a concrete type creates the array of values for storage:

```
──────────── Splat.cpp ────────────
   template<class E_> void SetScalar(const E_& e)
   {
       val_.Resize(1, 1);
       val_(0, 0) = e;
5  }
   // ...
   void operator=(double d) override { SetScalar(d); }
   // ...
```

and so on through the other concrete types.

A reference is a unique identifier for an already-written object, but our scheme requires us to form it for an object about to be written.

```
──────────── Splat.cpp ────────────
   bool StoreRef(const Storable_* object) override
   {
       auto ot = sharedTags_.find(object);
       if (ot != sharedTags_.end())
5      {
           SetScalar(ot->second);  // store string in lieu of tag
           return true;
       }
       auto tag = TAG_PREFACE + ToString(1 + sharedTags_.size());
10     sharedTags_.insert(make_pair(object, tag));
       SetTag(tag);
       return false;
   }
```

Thus failure to find a stored reference in **sharedTags_** causes a reference to be stored for the future. This dovetails with our requirement that objects must write their contents iff **StoreRef** returns **false**.

5.4.2 *Display Format*

To create the final output, we must decide on a format for the table. We use indentation to distinguish levels of nesting, so the only entry in the first column is the type of the (outermost) object; we prepend a special character (~) which will later be used to separate type strings from tags. The second column contains the child names, and each child's contents start in the third column. (If we were adding type information, it would occupy an additional column at each level of nesting.)

Child objects are written identically to the parent object, but indented by two columns; however, they are also tagged for possible reuse. We prepend, before the type string, a tag string created by **StoreRef** (and not containing ~) which is unique within the archive.

```
                              Splat.cpp
   void Write
       (Matrix_<Cell_>& dst,
           int row_offset,
           int col_offset)
5  const
   {
       if (!val_.Empty())
       {
           for (int ir = 0; ir < val_.Rows(); ++ir)
10         {
               const auto& row = val_.Row(ir);
               auto out = dst.Row(ir + row_offset).begin() + col_offset;
                   copy(row.begin(), row.end(), out);
           }
15         return;
       }
       // composite object, first cell is tag and type
       dst(row_offset, col_offset) = tag_ + OBJECT_PREFACE + type_;
       for (auto c : children_)
20     {
           dst(row_offset, 1 + col_offset) = c.first;
           c.second->Write(dst, row_offset, 2 + col_offset);
           row_offset += c.second->Rows();
       }
25 }
```

Each child states its size, letting us update `row_offset`; thus the rows allotted to each child begin with the row containing its name, and end at the next row with a nonempty entry in that column.

5.4.3 *Extraction*

The `Unsplat_` reader must decipher the table thus created, or more generally, a sub-matrix of the table. Before performing any actions, the reader will receive some type information from its own caller, which will invoke a type-specific function like `AsString` or else ask for the `Type` of a polymorphic object.

When reading the archive, we will check the upper left cell of each child for a type and tag. The child is itself an instance of `View_`, so we add member functions supporting these queries:[6]

---

[6]There is no distinction in this archive between object tags and literal strings, so a reader could request a string and obtain the string value of an object tag. This is harmless in practice, but could be avoided by using an additional column at each level of nesting to store distinguishing type information.

```
                    ─── Splat.cpp ───
     String_ Type() const override
     {
        const Cell_& c = data_(rowStart_, colStart_);
        if (c.type_ != Cell_::Type_::STRING)
5          return String_();
        auto pt = c.s_.find(OBJECT_PREFACE);
        if (pt == String_::npos)
           return String_();
        return c.s_.substr(pt + OBJECT_PREFACE.size());
10   }
     String_ Tag() const
     {
        const Cell_& c = data_(rowStart_, colStart_);
        if (c.type_ != Cell_::Type_::STRING)
15         return String_();
        if (c.s_.substr(0, TAG_PREFACE.size()) != TAG_PREFACE)
           return String_();
        auto pt = c.s_.find(OBJECT_PREFACE);
        return c.s_.substr(0, pt);
20   }
```

The different fundamental types are extracted by supporting functions which separate the handling of element type and of dimension. For example:

```
                    ─── Splat.cpp ───
     const Cell_& GetScalar() const
     {
        REQUIRE(rowStop_ == rowStart_ + 1,
           "Multi-line entry is not a scalar");
5      REQUIRE(colStart_ == data_.Cols() - 1 ||
           Cell::IsEmpty(data_(rowStart_, colStart_ + 1)),
           "Multi-row entry is not a scalar");
        return data_(rowStart_, colStart_);
     }
```

This tests a section of the displayed object to see if it might be a single number: if so, it will have one row, the remainder of which will be blank. (We would also check **dataType_** information at this point, had we chosen to use it.) This supports the member functions of **View_** which seek to interpret the stored value as a scalar, *i.e.* AsDouble:

```
                    ─── Splat.cpp ───
     double ExtractDouble(const Cell_& src) // nonmember
     {
        switch (src.type_)
        {
5      case Cell_::Type_::NUMBER:
           return src.d_;
        case Cell_::Type_::STRING:
```

```
            return String::ExtractDouble(src.s_);
        default:
10          THROW("Element is not a number");
        }
    }
    // ...
    double AsDouble() const override
15  {
        return ExtractDouble(GetScalar());
    }
```

Other **As** functions will be implemented along the same lines.

Similarly, we can find an iterator range for the contents of a vector (or a two-dimensional range for a matrix), then coerce each element to the requested type:

```
────────────── Splat.cpp ──────────────
    typedef Matrix_<Cell_>::Row_::const_iterator row_ci;
    pair<row_ci, row_ci> VectorRange() const
    {
        REQUIRE(rowStop_ == rowStart_ + 1,
5           "Multi-line entry is not a vector");
        int colStop = colStart_ + 1;
        while (colStop < data_.Cols() &&
            !Cell::IsEmpty(data_(rowStart_, colStop)))
            ++colStop;
10      auto row = data_.Row(rowStart_).begin();
        return make_pair(row + colStart_, row + colStop);
    }
    // ...
    Vector_<> AsDoubleVector() const override
15  {
        return TranslateRange(VectorRange(), ExtractDouble);
    }
```

To get a list of **Children**, we walk down the next-to-leftmost column (after checking that the leftmost column contains the expected type and reference information). Each nonblank entry in this column is the name of a child, whose data live in the corresponding submatrix.

```
────────────── Splat.cpp ──────────────
    const View_& Child(const String_& name)const override
    {
        NOTICE(name);
        assert(!Type().empty());    // atom has no children
5       const int nameCol = colStart_ + 1;
        assert(nameCol < data_.Cols());
        for (int ir = rowStart_; ir < rowStop_; ++ir)
        {
```

```
            if (data_(ir, nameCol) == name)
10          {
                // found it
                int jr = ir + 1;
                while (jr < rowStop_ &&
                  Cell::IsEmpty(data_(jr, nameCol)))
15                ++jr;
                children_.push_back(shared_ptr<XUnSplat_>
                (new XUnSplat_
                        (data_, ir, jr, nameCol + 1)));
                return *children_.back();
20          }
        }
        THROW("No such child");
    }
```

A child is itself a fully-fledged `View_`.

The query for an already-built object is deceptively simple:

Splat.cpp

```
    Handle_<Storable_>& Known(Archive::Built_& built)
    const override
    {
      return built.known_[Tag()];
5   }
```

The `Storable_` object thus referenced may or may not exist. In the latter case, we are relying on `Archive::Extract` to read the `View_`'s contents, build the object, and emplace it in the `Handle_&` we have provided.

A single `Built_` object is shared by reference among all the `View_s` (remember that a child is also a `View_`). To ensure that this reference remains valid, we define the derived `View_` class in a local namespace and access it only through the function:

Splat.cpp

```
Handle_<Storable_> UnSplat(const Matrix_<Cell_>& src)
{
    XUnSplat_ task(src, 0, src.Rows(), 0);
    NOTE("Extracting object from splatted data");
5   Archive::Built_ built;
    return Archive::Extract(task, built);
}
```

5.4.4 Refinements

The above describes a minimal object display engine; several refinements will likely be desirable.

- Allow specification of an *element path* so that a sub-object, rather than the entire object, can be viewed.
- Add a tab-separated file reader/writer to convert the `Matrix_` data to files; or generalize the data type used to support both matrices and files.
- Allow truncation of the output at a user-input tree depth, so that objects can be inspected gradually from the top.
- Interact with the repository, replacing child data (beyond the specified depth) with a repository handle (which can be splatted in turn, if desired).

Since the table output from `Splat` can be written to a file, it can serve as the sole archive method for many purposes. However, there is a widespread feeling that JSON files, or some similar format with third-party backing, are more "official" and more suited for books and records.

5.5 Auditing

During a long computation, we will form and discard complicated objects; for example, a risk run usually involves the creation of temporary bumped models. Inspection of these objects can help greatly in debugging or validating a process; but they are no longer with us.

We would like to have the option to preserve these objects. Clearly they cannot be carried out along the call stack.[7] But we can store them in our environment.

5.5.1 *Bag*

The first requirement is a place to store objects. We will require that they be `Storable_`, since a non-storable object is of little utility outside its local context; it cannot be put in the repository, or displayed, or written to a file. We need a key facility to attach additional information and to let us search for a given piece; but these keys might not be unique. Thus we arrive at

Bag.h

```
struct Bag_ : Storable_
{
    typedef multimap<String_, Handle_<Storable_>> map_t;
    map_t contents_;
    Bag_(const String_& nm, contents_) : Storable_("Bag", nm) {}
    void Write(Archive::Store_& dst) const override;
};
```

---

[7]It is an interesting exercise to enumerate the distinct ways in which this is infeasible.

The public interface of `Bag_` must allow us to view, insert and erase objects; thus we simply make `vals_` public data. A `Bag_` is very floppy, and can be manipulated at will (*e.g.*, by the `Auditor_` creating it), but a `Handle_<Bag_>` has fixed contents.

5.5.2 *Filling Up*

We need an auditor, an object which can be passed within the environment. Functions which create temporary `Storable_` objects – *i.e.*, not their return values – can and generally should put them in handles and show them to the auditor. We may go out of our way to make temporary objects `Storable_` for this purpose.

The auditor may also be used to communicate between successive iterations of a complex algorithm (*e.g.*, between base and bumped valuations). Thus we provide a query interface to check for the presence of some named object.

```
_____ Audit.h _____
   class Auditor_ : public Environment_::Entry_
   {
   public:
      virtual void Notice
5         (const String_& key,
            const Handle_<Storable_>& value)
      const = 0;

      virtual Vector_<Handle_<Storable_>> Find
10        (const String_& key)
      const = 0;
   };
```

We will support these with **namespace Environment** utilities[8]

```
_____ Audit.h _____
   template<class T_> void Audit(_ENV, const String_& key,
         const Handle_<T_>& value)
   {
      AuditBase(_env, key, handle_cast<Storable_>(value));
5  }
```

---

[8]These are implemented as two separate functions so that `Environment::Iterate` will be instantiated only once.

```
                    ──────────── Audit.cpp ────────────
void Environment::AuditBase(_ENV,
    const String_& key, const Handle_<Storable_>& value)
{

    ShowToAuditor_ f(key, handle_cast<Storable_>(value));
5   Environment::Iterate(_env, f);
}
```

where **ShowToAuditor_** is a function object which checks (using **dynamic_cast**) whether the input **Environment::Entry_** is an **Auditor_**, and if so calls **Notice**. The names must be chosen to show the direction of information flow, so we avoid words like "remember" – which means either to store in, or to recall from, memory – and "show," since it is not clear whether an object is being shown to or by the auditor.[9]

User code, once it has a handle **localThing** to whatever object it creates, simply calls

```
Environment::Audit(_env, "name", localThing);
```

which gives any auditor in the environment the chance to **Notice** it.

The common use case of **Find** is to fetch forth a single previously **Noticed** object, and place it in a user-provided handle:

```
                    ──────────── Audit.h ────────────
template<class T_> struct Recall_
{
    const String_ key_;
    Handle_<T_>* value_;
5   const Environment_* env_;
    Recall_(_ENV, const String_& key, Handle_<T_>* value)
        : env_(_env), key_(key), value_(value) {}
    void operator()(const Entry_& env) const
    {
10      if (auto audit = dynamic_cast<const Auditor_*>(&env))
        {
            auto fh = audit->Find(key_);
            for (auto h : fh)
            {
15              Handle_<T_> temp = handle_cast<T_>(h);
                if (!temp.Empty() && temp != *value_)
                {
                    REQUIRE(value_->Empty(),
                     "Conflicting recollections");
20                  *value_ = temp;
                }
```

────────────────────────

[9]The similarity of this name to our NOTICE macro is largely deliberate, and there is no chance of confusing the two.

```
              }
          }
      }
25 };

   template<class T_> void Recall
      (_ENV, const String_& key, Handle_<T_>* value)
   {
30    assert(value && value->Empty());
      NOTICE(key);
      Iterate(_env, Recall_<T_>(_env, key, value));
   }
```

This somewhat awkward code supports a reasonably compact usage idiom:

```
   Handle_<MyType_> local;
   Environment::Recall(_env, "theKey", &local);
   if (local.Empty())
   {
5     // make it ourselves
      // ...
      Environment::Audit(_env, "theKey", local);
   }
```

Sec. 14.7.1 shows a realistic example. This provides a context with which a computation can detect that it is a bumped case (a perturbation), and have some chosen data available from the base computation. We can use this as a generic stabilization method for perturbations; see Sec. 14.4.

5.5.3 *Audit Types*

Auditors are not obliged to show all their contents to any caller of `Find`; also note that we provide no facility to find all the keys. The expected implementation of `Auditor_` holds a `Bag` into which it might place objects:

```
                    ─── Audit.cpp ───
   struct AuditorImp_ : Auditor_
   {
      shared_ptr<Bag_> mine_;
      enum
5     {
         PASSIVE,
         READING,
         READING_EXCLUSIVE,   // avoid vast memory use
         SHOWING,
10    } mode_;
      // ...
   };
```

The default constructor will set the `mode_` to `READING`, meaning that the auditor should hang onto every object sent to `Notice`.

```cpp
                      ── Audit.cpp ──
void AuditorImp_::Notice
   (const String_& key,
    const Handle_<Storable_>& value)
const
{
    switch (mode_)
    {
    case READING_EXCLUSIVE:
        mine_->contents_.erase(key);   // and fall through
    case READING:
        mine_->contents_.insert(make_pair(key, value));
        break;
    }
}
```

Other modes ignore those objects but will `Find` their stored objects:

```cpp
                      ── Audit.cpp ──
Vector_<Handle_<Storable_>> AuditorImp_::Find
   (const String_& key)
const
{
    static Get2nd_<String_, Handle_<Storable_>> getV;
    Vector_<Handle_<Storable_>> retval;
    if (mode_ == SHOWING)
    {
        auto range = mine_->vals_.equal_range(key);
        transform(range.first, range.second,
            back_inserter(retval), getV);
    }
    return retval;
}
```

An `AuditorImp_` belongs to its creator, who alone can access its full data (except by sneaky casting, easily found by a global code search) or change its mode.

To audit a given function, we add an empty `AuditorImp_` to the environment before the call, and later pick up its `Bag_` of data. For a toy example, consider the helper function `Interp1_Get` from Sec. 3.2; we can write

```
Handle_<Bag_> Interp1_Get_Audit(_ENV,
    const Interp1_& f, const Vector_<>& x, Vector_<>* y)
{
    AuditorImp_ audit;
    ENV_ADD(audit);
    Interp1_Get(_env, f, x, y);
    return audit.mine_;
}
```

Of course, this will accomplish nothing unless the specific `Interp1_` implementation has chosen to call `Environment::Audit`, which is unlikely for a linear or spline interpolator.

We might also create special-purpose auditors which ignore objects unless, *e.g.*, their keys match some pattern. A more general auditing mechanism is best implemented as part of a larger project, creation of an analytics scripting language, which is outside the scope of this volume.

5.6   More on Repositories

5.6.1   *Naming*

To store objects in our in-process repository, we must decide how to name them; *i.e.*, to specify how "long names", which will be used as repository keys, will be generated for a given object. Each `Storable_` has a `type_` and `name_`, which should both be included in the long name; but we must also provide a way to distinguish successive instances with the same type and name (*e.g.*, a yield curve which is continually updated with new market data). Of course, the long name must be unique within an interactive session; thus we must attach extra text within the long name. Several approaches are possible:

- A version number or "ticker", incremented each time a new object is created;
- Variants where a separate ticker is used for each type, or each name, or each combination thereof;
- A string representation of the object's address in memory;
- A GUID (128 bits, 32 hex chars, 20-26 printable chars);
- The chip-independent part of a GUID ($\frac{5}{8}$ as big).

The first two approaches seem more friendly to the user, who sees each object with a numeric version rather than an unreadable multi-character "dongle." However, this is actually a disadvantage if it invites the user to type long names by hand, thus breaking the dependency chain of a calculation. We prefer the third approach, which can be implemented with about seven characters of extra information.

Long names should also be immediately identifiable as such. For this purpose it is worthwhile to reserve a character like ~ or &, which will not naturally be used in object naming, and use it as a separator; thus long names will look like ~YC~USD1~g8h6w04.

5.6.2 *Matching*

Fetching forth an object's long name from incomplete information is not always a mistake. For instance, a pricing spreadsheet may be intended to exist independently of any single yield curve, and to operate using whatever curve (with the appropriate currency) is present in the environment. Thus we have `Repository::Find` to search the repository and return long names.

This is best done with *pattern matching*. Rather than just supply the beginning of the name, the user gives a pattern to match, and we return all long names matching it. So matching ~YC~ is a search for all yield curves,[10] while matching ~YC~USD1~ is a search for all yield curves with that particular name. Since C++11 includes standard library regular expressions, using anything else would smack of masochism.

Users will control the names given, and can adapt their searches to their own naming conventions. A couple of refinements are useful:

- Simultaneous match of multiple patterns, both "and" and "or".
- An optional flag to enforce uniqueness, returning an error unless exactly one match is found.
- Permitting ^ to consume the leading ~ or other special character.

This function lets users examine the repository, to find an object or to fetch all objects (*e.g.*, for creating a bag of environment).

5.6.3 *Capturing State*

Within a user session in most interactive languages (*e.g.*, Mathematica or Python) the program state evolves. The interpreter state or kernel,

---

[10] And also, unless we change our naming system, for other objects which happen to be named "YC".

a global store of definitions, is updated as new assignments are made.[11] Excel has the peculiar property that there is no kernel; or, more precisely, the spreadsheet explicitly holds all the definitions.

Adding a repository of opaque objects breaks Excel's kernel-free paradigm. Many of our spreadsheets will have dependencies, created by the use of Repository.Find, which are hidden from Excel. Thus different users can see different results, and debugging sessions cannot always reproduce the problems seen in production.

We can minimize the damage by providing a convenient way to capture the entire state of an interactive session. Since Excel is an important computational platform in finance, it is worth designing our library to allow this. This leads us to one important design constraint: *the repository must be the only form of state which persists between calls to public functions*. The user can

- Interactively list all repository objects;
- Create a Bag_ from a list of objects; and
- Serialize and save that Bag_.

Thus already it is easy to permanently store the repository state: A later session reads this Bag_ of state and views its contents, which causes the contained handles to be added the that session's repository; see Sec. 5.5.1. We need to ensure that no global state evades this simple process.

5.6.4 *Unique Objects*

We choose to make the distinction between past and future part of the environment, rather than provide it as an explicit function input, since all valuation and risk functions need this information. This distinction is discussed in detail in Sec. 10.3; here we consider the storage of the *accounting date* which separates future payments from those already paid.

Some internal risk analyses, or payment and expiry reports, will be implemented by re-evaluating the portfolio after perturbing such a date. But the user may also override those dates, *e.g.*, to run scenarios of the past or future. Thus they must be captured as persistent global state; as we have just seen, this is best accomplished by using the object repository to hold them.

---

[11]This global store is what is lost when restarting the kernel in a Mathematica or iPython notebook.

At the same time, the repository is somewhat Excel-specific, and we wish to prevent its use within analytic code. (It is possible and sometimes tempting to use the repository as a crude auditor; this leads to brittle code which is hard to test or debug.) We emphasize this in our source code with the file name _Repository.h; the leading underscore denotes an interface-level rather than an internal header.

To resolve this level conflict, we create a low-level abstract storage mechanism:

```
─────────────── Globals.h ───────────────
   namespace Global
   {
      class Store_ : noncopyable
      {
 5    public:
         virtual ~Store_();
         virtual void Set(const String_& name, const Cell_& value)
         = 0;
         virtual Cell_ Get(const String_& name) = 0;
10    };

      Store_& TheDataStore();
      void SetTheDataStore(Store_* orphan);    // we take over the
                                                      memory
15 }
```

The repository code will instantiate a concrete instance derived from Store_ and SetTheDataStore to use it; but clients seeking a global date need not be aware of these details. Internal analytics code away from the user interface cannot justifiably access the repository's table of user-level handles, so they work with only the more restricted Global::Store_ interface.

Upon initialization of a session, the repository will be empty; in particular it will not contain any system dates. We query and initialize these dates through interfaces which in turn access a common utility function.[12]

```
─────────────── Globals.cpp ───────────────
   Date_ GetGlobalDate(const String_& which)
   {
      LOCK_STORE;
      Cell_ stored = Global::TheDataStore().Get(which);
 5    if (Cell::IsEmpty(stored))
      {
         // no global date set; initialize to system date
         int yy, mm, dd;
         Host::LocalTime(&yy, &mm, &dd);
```

───────────────────────────────

[12]We might choose to initialize based on the load time, rather than the call time as shown here.

```
10      stored = Date_(yy, mm, dd);
        Global::TheDataStore().Set(which, stored);
     }
     return Cell::ToDate(stored);
 }
```

Remember that `TheDataStore()` interacts with the repository, where the
object will be kept (as a `Cell_` in a storable `Box_`). This method initializes
the accounting date (or other date) to the local date when it is first called.
In a global 24/5 operation, this might not be adequate, but the solution
can only be to always explicitly initialize the global dates.

Now we set the accounting date by storing a new object, which will
overwrite any others already in the repository. This is wrapped in functions
in a deliberately glaring namespace:

———————————————— Globals.h ————————————————

```
namespace XGLOBAL
{
    void SetAccountingDate(const Date_& dt);
    ScopedOverride_<Date_> SetAccountingDateInScope
 5       (const Date_& dt);
}
```

The latter function allows temporary manipulation of the global using the
RAII idiom.

Note that the `XGLOBAL` functions are write-only; the saved date is
`private` in `ScopedOverride_`. Read access is thus channeled through the
`Environment`.

Now there is an approved way to read a global date from the environment, presuming that our caller has already supplied the `Global::Dates_`
accessor using `ENV_ADD_TYPE` or `ENV_SEED_TYPE`:

```
Environment::Find<Global::Dates_>(_env)->AccountingDate();
```

Unfortunately, the code to accomplish the same thing locally, while also
hiding our intention from our caller, is shorter and clearer:

```
Global::Dates_().AccountingDate();
```

To prevent this would require giving `Global::Dates_` a private constructor, making `Environment_` its friend, and then adding
a special-purpose `Environment_` function to be used in place of
`ENV_ADD_TYPE(Global::Dates_)`. This seems excessive, and I have not
implemented it.

Chapter 6

Testing Framework

Like any large software project, our library will evolve through time and will need constant upkeep to remain reliable. In attempting to do more tasks with less code, we are inevitably forcing code to be used in many different contexts, so a bug will often only be manifest in some uncommon or unexpected use case. Also, testing of the interaction between high-level components requires a great deal of low-level work, even if the low-level components are known to be reliable.

6.1 Component Tests

Inside the code, we can add tests of individual functions or classes. These tests are themselves functions, which can in theory be the **main** functions of many individual executable tests.[1] This decreases the overhead of selecting a specific test, at the expense of increased maintenance and link time. We prefer to run all tests automatically, rather than rely on the developer's judgement of which are needed. This leads us to the opposite approach: a single large test executable, which calls separately written test functions in sequence. We will provide the test executable with a singleton registry of functions to call, and use our usual registration idiom:

```
―――――――――――――――― Test.h ――――――――――――――――
namespace Test
{
    typedef void(*func_t)();
    void Register(func_t func);
    void Fail(const char* msg);
    void Fail(const String_& msg);
}
```

―――――――――――――――――――――――――――――――――――――
[1] This approach seems to be implicit in Lakos's *Large-Scale C++ Software Design*.

```
     struct ComponentTest_
10   {
        ComponentTest_(Test::func_t fn) { Test::Register(fn); }
     };
     #define TESTFUNC(nm) \
     void nm();              \
15   static const ComponentTest_ test_register__##nm(nm); \
     void nm()
```

This definition of TESTFUNC lets us use it as a function signature directly
in the code:

─────────────────── *NCDF.test.cpp* ───────────────────
```
     TESTFUNC(TestNCDF)
     {
        using SpecialFunctions::N;
        TEST(N(0.0) == 0.5);
5    // ...
```

Here we have introduced a new macro, which we implement as:

─────────────────── *Test.h* ───────────────────
```
     #define TEST(cond) if (cond); else Test::Fail(#cond)
```

The function Test::Fail implements the consequences of failure; most
likely it will simply display a message to cout and increment an error
count, but more sophisticated behaviors are possible.

Since a large part of our work is numerical, we will implement some
other macros to streamline numerical tests.

─────────────────── *Test.h* ───────────────────
```
     #define TEST_SIMEQ(x, y, eps)    \
     {const double myX = x, myY = y;        \
     if (fabs(myX - myY)         \
          <= eps * Max(1.0, Max(fabs(myX), fabs(myY)))); \
5    else Test::Fail(#x "~" #y "(within " #eps ")")

     #define TEST_EQ(x, y) TEST_SIMEQ(x, y, DA::EPSILON)
```

The input tolerance eps is thus interpreted as a relative tolerance for
large numbers, and as an absolute tolerance for small numbers. This is
the test we most often desire; we can easily implement other macros like
TEST_ABSOLUTE and TEST_RELATIVE as needed.

Another macro tests for the emission of an error message:

─────────────────── *Test.h* ───────────────────
```
     #define TEST_ERROR(x, m) \
     try {x; Test::Fail(#x " did not generate the error");} \
     catch (Exception_& e) {TEST(e.Display() == m);}
```

6.1.1 *Physical Structure*

We are reluctant to put tests directly into a source file, since they clutter the implementation. But the test code often needs access to the full implementation of a class, not just the public interface; and keeping the implementation insulated in the source file is itself an important design aim.

Our solution is to include *the source itself* within a testing file. For example, we might create NCDF.test.cpp which would begin

```
———————————— NCDF.test.cpp ————————————
#include "NCDF.cpp"
#include "ComponentTest.h"

TESTFUNC(TestNCDF)
```

and go on to implement other testing functions.

We then create two separate builds, incorporating either NCDF.test.cpp or NCDF.cpp; these build the library with or without component tests. The latter is more suitable for release into production.

The test executable implements Test::Register – thus, during the loading of the library (with component tests), each test declared using TESTFUNC will be registered. If we wish to load the library with component tests into a different executable, we must provide a stub version of Test::Register.

6.1.2 *Reuse*

Creating a high-level object for testing requires many prerequisite objects: for instance, to test a risk computation, we need a model, which needs a yield curve, which needs instruments. This threatens to make testing code cumbersome to the point of uselessness.

We must add additional functionality, in the test files, to create lower-level objects which can be reused when testing high-level objects. It would be nice to reuse the auditing tools from Sec. 5.5 to store these objects, rather than building them from scratch for each test; unfortunately, that would require tracking the dependencies between component tests. Thus we stick with the low-tech solution of simply defining helper functions and declaring them in test header files for later use. A typical example is:

```
———————————— YcBuild.test.h ————————————
Handle_<YieldCurve_> Test::YC_GBP30Y(const Date_& quote_date)
```

6.2   Regression Tests

Tests which use only the public interface of the library need not be written
in C++. We can take advantage of interactive environments, notably Excel,
to more quickly put together such a test. These are more useful for regression testing, to flag unexpected changes, than for component correctness
testing.

Sec. 5.6 describes how we can "bag up" the whole environment of an
interactive session; thus a stored regression test can consist of a bag of
objects, plus a continuation of the interactive session. For example, we
might store a file containing the session state, and an Excel spreadsheet
to be run after the file is loaded. The spreadsheet's cells, or some subset
thereof (*e.g.*, those marked with a cell comment), could be compared across
releases.

6.2.1   *Repository Instrumentation*

It is not necessary to bag the whole environment; if we mediate repository access, we can add code to record which objects were fetched
during a computation. This process is known as *instrumentation*, in
this case within `Environment::Fetch`. We rename the simple version
`ObjectAccess_::Fetch` from Sec. 3.7.2 to `XFetch`, to highlight that it
should not be called directly, and write:

```
──────────────── EnvironmentRepository.h ────────────────
     template<class T_> Handle_<T_> Fetch
        (_ENV, const String_& tag, bool opt = false)
     {
        NOTICE("Handle tag", tag);
5       Handle_<Storable_> base = FetchBase(_env, tag, opt);
        Handle_<T_> retval = HandleCast<T_>(base);
        Require(_env, !retval.Empty() || base.Empty(),
              "Stored object is of wrong type");
        return retval;
10   }
```

```
──────────────── EnvironmentRepository.cpp ────────────────
   Handle_<Storable_> Environment::FetchBase
      (_ENV, const String_& tag, bool opt)
   {
      auto access = Extract1<ObjectAccess_>(_env);
5     Require(_env, opt || access, "No repository access");
      if (!access)
         return Handle_<Storable_>();
      else if (RepositoryListener_* listen = TheListener())
```

```
      return (*listen)(_env, *access, tag, opt);
10  else
      return access->XFetch<Storable_>(tag, opt);
}
```

This requires a new class **RepositoryListener_**, stored as a singleton and accessed as **TheListener**.

One such listener simply records the handles returned:

EnvironmentRepository.cpp
```
struct RepositoryRecording_ : RepositoryListener_
{
    map<String_, Handle_<Storable_>> seen_;
    Handle_<Storable_> operator()
5       (_ENV, const ObjectAccess_& access,
        const String_& tag, bool optional)
    {
        Handle_<Storable_> retval
            = access.Fetch<Storable_>(tag, optional);
10      seen_[tag] = retval;
        return retval;
    }
};
```

To "record" part of a session now means the following:

(1) Create an instance of **RepositoryRecording_** and set **TheListener** to it.
(2) Run the desired part of the session.
(3) Reset **TheListener** to 0.

After this process, the recording will have been populated with all the repository queries made.

In practice, repository instrumentation works best in combination with a scripting language, making it simple to turn a single execution of a script into a self-contained regression test.

6.3   No Silver Bullet

The ongoing challenge, of course, is to keep both component and regression tests up-to-date and comprehensive. Alas, this is still a matter of discipline, with no technological silver bullets on offer.

This page intentionally left blank

Chapter 7

Further Maths

Besides the vector and matrix computations discussed in Ch. 4, we require a good deal of other mathematical functionality, some of it quite specialized to the field of derivatives.

7.1 Interpolation

Our approach to construction of (one-dimensional) interpolants is unsurprising: we create an abstract base class `Interp1_` and derive concrete classes from it. Our only innovation is in cubic splines, where we support the ability to set the first, second or third derivative at each boundary. In namespace Interp, we write:

```
──────────────── InterpCubic.h ────────────────
    struct Boundary_
    {
        int order_;
        double value_;
5       Boundary_(order_, value_) {]
    };
```

Thus `Boundary_(3, 0.0)` instructs the decomposition to set the third derivative to zero (I like to refer to this as the "super-natural" spline). We set the left and right boundaries independently when splining:

```
──────────────── InterpCubic.h ────────────────
    Interp1_* NewCubic
        (const String_& name,
         const Vector_<>& x,
         const Vector_<>& f,
5        const Boundary_& lhs,
         const Boundary_& rhs);
```

103

The factory function calls a constructor of a *local class* defined only within
the source file. The code for this is based on the **spline** routine of *Numerical Recipes*, but we have changed the treatment of boundaries to support
our wider interface, and also interchanged y2 and u to remove a gratuitous
relabeling.

```
——————————— InterpCubic.cpp ———————————
Cubic1_::Cubic1_
   (const String_& name,
    x_,
    const Vector_<>& f,
    const Interp::Boundary_& lhs,
    const Interp::Boundary_& rhs)
   :
Interp1_(name),
f_(f),
fpp_(f.size())
{
    assert(x_.size() > 2 && IsMonotonic(x_));
    assert(x_.size() == f_.size());
    const int n = x_.size();
    Vector_<> u(n - 1);
    switch (lhs.order_)    // set left boundary
    {
    default:
        assert(!"Invalid boundary order");
    case 1:
        {
            const double dx = x_[1] - x_[0];
            fpp_[0] = ((f_[1] - f_[0]) / dx - lhs.value_) * 3.0 / dx;
            u[0] = -0.5;
        }
        break;
    case 2:
        fpp_[0] = lhs.value_;
        u[0] = 0.0;
        break;
    case 3:
        fpp_[0] = -(x_[1] - x_[0]) * lhs.value_;
        u[0] = 1.0;
        break;
    }
    for (int i = 1; i < n - 1; ++i)    // decomposition
    {
        const double dx = x_[i] - x_[i - 1];
        const double d2 = x_[i + 1] - x_[i - 1];
        const double sig = dx / d2;
        const double p = sig * u[i - 1] + 2.0;
        u[i] = (sig - 1.0) / p;
        const double temp = (f_[i+1] - f_[i]) / (x_[i+1] - x_[i])
```

```
                       - (f_[i] - f_[i-1]) / dx;
45          fpp_[i] = (6.0 * temp - dx * fpp_[i - 1]) / (p * d2);
         }
         switch (rhs.order_)    // set right boundary
         {
         default:
50           assert(!"Invalid boundary order");
         case 1:
             {
                 const double dx = x_[n - 1] - x_[n - 2];
                 const double un = (3.0 / dx) *
55                   (rhs.value_ - (f_[n - 1] - f_[n - 2]) / dx);
                 fpp_[n - 1] = (2.0 * un - fpp_[n-2]) / (2.0 + u[n-2]);
             }
             break;
         case 2:
60           fpp_[n - 1] = rhs.value_;
             break;
         case 3:
             fpp_[n - 1] = ((x_[n-1]-x_[n-2]) * rhs.value_ + fpp_[n-2])
                 / (1.0 - u[n - 2]);
65           break;
         }
         for (int k = n - 2; k >= 0; --k)  // backsubstitution
             fpp_[k] += u[k] * fpp_[k + 1];
}
```

7.1.1 *Functions of Time*

The x-variable of an interpolation may be something other than a **double**, most frequently a **DateTime_**. We support this, for any type convertible to **double**, by wrapping our existing interpolators; *e.g.*, in one dimension we have:

```
                           ──── Interp.h ────
   namespace Interp1
   {
       template<class T_> struct Of_
       {
5          Handle_<Interp1> imp_;
           Of_(const Handle_<Interp1>& imp) : imp_(imp) {}
           double operator()(const T_& x) const
               { return (*imp_)(NumericValueOf(x)); }
       };
10 }
```

NumericValueOf is a trait of the type of x, which the compiler finds by overload resolution. In the case of **DateTime_**, we provide

```
————————— DateTime.h —————————
inline double NumericValueOf(const DateTime_& src)
  { return NumericValueOf(src.Date()) + src.Frac(); }
```

7.2 Special Functions

7.2.1 *The Normal Distribution*

The most important special functions in finance are the normal cumulative distribution $N(x)$ and its inverse.[1] The most widespread public-domain implementation is that of Abramowitz and Stegun from their *Handbook of Mathematical Functions*; but it is not terribly accurate, and several shops have built improved in-house approximations.

Before committing to this approach, we must distinguish between cases where high accuracy is needed, and those where speed is more important. It is quite likely that we could run an entire real-world operation using only a low-accuracy approximation – say, to four decimal places – of N; this would be tantamount to using a slightly different distribution in pricing. The main difficulty would arise from lack of closure under convolution.[2]

Given an approximate value for $N^{-1}(x)$, such as the output of a low-precision analytical approximation, we can radically improve it by a single "polishing" using Newton's method. Suppose $y \sim N^{-1}(x)$, and let $y' \equiv y + (x - N(y))/\phi(y)$ where ϕ is the normal density. A single iteration usually suffices to achieve double-precision accuracy. Thus at the expense of one evaluation of N and one of ϕ, we can enforce consistency between our two approximations.

This leads us to the interface, in namespace SpecialFunctions,

```
————————— SpecialFunctions.h —————————
double NCDF(double z, bool precise = true);
double InverseNCDF
       (double x, bool precise = true, bool polish = true);
```

The defaults are chosen for reliability rather than speed. This is a good general practice: in known hotspots, or after profiling, we can switch over to the less-precise fast method, but we should never set it as the default.

Inside these functions, we will have a top-level switch based on the precise flag; the computational cost of this is insignificant. One candidate

---

[1] Criticism of models based on normal distributions is widespread and justified, but these distributions also supply the foundation for more realistic models.

[2] There is no pressing reason to try this experiment.

for an approximate NCDF or InverseNCDF is cubic spline on a precomputed set of points; it is possible to get absolute errors in N under 10^{-6} everywhere with as few as 16 spline points.[3]

```
———————————— SpecialFunctions.cpp ————————————
static const double MIN_SPLINE_X = -3.734582185;
static const double MIN_SPLINE_F = 9.47235E-05;
Interp1_* MakeNcdfSpline()
{
5       static const Vector_<> x =
            { MIN_SPLINE_X, -3.347382781, -3.030883722, -2.75090681,
                -2.492289824, -2.243141537,    -1.992179668, -1.494029881,
                  -1.290815576, -1.120050999, -0.954303629, -0.792072249,
                -0.629093487, -0.460389924, -0.276889742, 0.0 };
10      static const Vector_<> f =
            { MIN_SPLINE_F, 0.000408582, 0.001219907, 0.002972237,
                0.00634685, 0.012444548, 0.023176395, 0.067583453,
                  0.098383227, 0.131345731, 0.16996458, 0.214158839,
                0.264643073, 0.322617682, 0.39093184, 0.5 };
15      const Interp::Boundary_ lhs(1, 0.000373538);
        const Interp::Boundary_ rhs(1, 0.39898679); // at x=0
        return Interp::NewCubic(String_(), x, f, lhs, rhs);
}
```

We will call this once to create a static spline interpolant:

```
———————————— NCDF.cpp ————————————
double NcdfBySpline(double z)
{
    static const scoped_ptr<Interp1_> SPLINE(MakeNcdfSpline());
    if (z > 0.0)
5       return 1.0 - NcdfBySpline(-z);
    return z > MIN_SPLINE_X
        ? (*SPLINE)(z)
        : MIN_SPLINE_F * exp(-1.1180061 *
            (Square(z) - Square(MIN_SPLINE_X)));
10 }
```

We choose to compute $N(x)$ at negative x in order to obtain accurate values when N is very small – even when $1 - N$ is indistinguishable from 1.

7.3 Root Solvers

There is a canonical way to write a rootfinder in C++, as a class with a virtual member which evaluates the objective function:

---

[3]In fact, this is one of the few sound applications of the cubic spline.

```
   class RootSearch_    // deprecated interface
   {
      // data representing state of the root search
   public:
 5    virtual ~RootSearch_();
      virtual double F(double x) const = 0;
      double Solve(double x_0, double target, double tol);
      // more Solve functions or fine-grained initializers
   // ...
10 };
```

Rootfinders are written this way because the C++ virtual function replaces the C callback, which in turn replaced the Fortran 77 call of func. Besides this legacy, the method has little to recommend it. The derived class implementing F is a pointless blemish, and separates the evaluation code from the calling code so that the flow of control is diverted through three different classes.

The problem is that the interface is inside out: rather than *providing* a service, the rootfinder *demands* one (*i.e.*, an implementation of F). To correct this, we write instead:

```
───────────────── Rootfind.h ─────────────────
   class Rootfinder_
   {
   public:
      virtual ~Rootfinder_();
 5    virtual double NextX() = 0;
      virtual void PutY(double y) = 0;
      virtual double BracketWidth() const = 0;
   };
```

A root search algorithm, like Ridders's or Brent's method, will be implemented in a concrete derived Rootfinder_. These methods allow the narrowing of a bracketing interval known to contain the root. The local variables of Solve in the callback-based implementation, which represent the state of the root search, become member data of the rootfinder:

```
───────────────── Brent.h ─────────────────
   class BracketedBrent_ : public Rootfinder_
   {
      pair<double, double> a_, b_, c_;
      const double tol_;
 5    bool bisect_;
      double d_;

      friend class Brent_;
```

```
        BracketedBrent_(tol_) {}      // uninitialized state
10      void Initialize
            (const pair<double, double>& low,
             const pair<double, double>& high);
    public:
        BracketedBrent_
15          (const pair<double, double>& low,
             const pair<double, double>& high,
             tol_);

20      double NextX() override;
        void PutY(double y) override;
        double BracketWidth() const override
        { return fabs(a_.first - b_.first); }
    };
```

The above is based on the routine **zbrent** from *Numerical Recipes*; we have combined the two variables **a** and **fa**, which describe an inverse quadratic interpolation point, into the pair **a_**, and likewise for **b_** and **c_**.

Software engineers will at this point wish to insert some checking code to ensure that **PutY** is never called twice in succession without an intervening **NextX**, and that **NextX** is idempotent (or likewise is never called twice). This is not difficult but also not valuable; the calling code is sufficiently simple that such errors just do not occur.

This rootfinder is still demanding a service from its client: it insists that the root be bracketed before it will do its work. To remedy this, we support an expanding "hunt" for a bracketing interval based on a single initial point:

```
——————————— Brent.h ———————————
    class Brent_ : public Rootfinder_
    {
        BracketedBrent_ engine_;
        // state for a non-bracketed rootfinder
5       enum class Phase_ { INITIALIZE, HUNT, BRACKETED } phase_;
        bool increasing_;   // guide direction of hunt
        double stepSize_, trialX_;
        pair<double, double> knownPoint_;
    public:
10      Brent_
            (double guess, double tolerance = DA::EPSILON,
             double step_size = 0.0);
        double NextX() override;
        void PutY(double y) override;
15      double BracketWidth() const override;
    };
```

The extra `INITIALIZE` state lets us create the rootfinder with just an
x-value, moving all the function evaluations inside the root search loop.
When the hunt has succeeded in bracketing a root, we use our `friend` status to initialize the Brent engine (during our `PutY`).[4] This lets us hold the
`engine_` data on the stack, saving a memory allocation.

Note that we favor the more complex Brent's method, rather than Ridders's. In our experience the latter takes on average 15–20% more function
evaluations; thus the one-time investment in coding Brent's method results
in a noticeable efficiency gain across many applications.

Now calling code becomes simple and transparent. We can make it still
simpler by creating a convergence-checking utility:

```
―――――――――――――― Rootfind.h ――――――――
struct Converged_
{
    double xtol_, ftol_;
    Converged_(xtol_, ftol_) {}
5   bool operator()(Rootfinder_& t, double e) const
    {
        t.PutY(e);
        return fabs(e) < ftol_ || t.BracketWidth() < xtol_;
    }
10 };
```

For example, a bootstrapped yield curve fitter might contain this inner
loop:

```
Brent_ task(guess);
for (int i = 0; ; ++i)
{
    REQUIRE(_env, i < ITERATION_LIMIT,
5       "Exhausted iterations in rootfind");
    values_->Back() = task.NextX();   // set our rate
    const double rate = target.ImpliedRate(*this);
    if (CONVERGED(task, rate - target_rate))
        break;
10 }
```

Here `CONVERGED` is a `static const` object embedding the rootfinder tolerances.

---

[4]Those allergic to friends can achieve almost the same effect by using placement `new` to construct the engine.

7.4 Underdetermined Search

But why should we use a bootstrapped yield curve fitter? Even the simplest yield curve has thousands of degrees of freedom – e.g., the one-day forward rates for each business day – and at most a few dozen market constraints. The customary bootstrapping approach is a Procrustean bed upon which "extra" degrees of freedom are lopped away until a solution is uniquely defined. But this truncation process is itself largely arbitrary, so the resulting curve is a mass of interpolation artifacts.

Unfortunately, not only has this lesson not been learned, but the same flawed methods have been propagated to other settings such as volatility calibration.

Consider the following problem of calibration, with yield curve fitting as a concrete example. We have n equality constraints to be matched, and $M \gg n$ commensurable degrees of freedom, like the forward rates over short non-overlapping intervals.[5] We write the constraints as $\vec{f}(\vec{x}) = 0$ where $\vec{x}$ is an M-dimensional vector representing a point in the parameter space. In the non-degenerate case, there is an $(M - n)$-dimensional manifold of solutions. Algorithms such as bootstrapping represent recipes for picking one point on this manifold (by restricting the solution search to an n-dimensional subspace); these recipes are clearly optimized for ease of implementation, not for the quality of the output.

To pick a particular solution less arbitrarily, we must define a figure of merit – a criterion for comparing the desirability of two competing solutions. This is a general optimization problem subject to nonlinear constraints, which will be computationally impractical for even medium-sized problems. But one widely studied subclass of problems can be solved quite efficiently: the *quadratic programming* problems in which the constraints are linear and the figure of merit is a positive definite quadratic form.[6] While this problem is not quite the one we face, it is worth understanding its solution in detail.

A quadratic form of $\vec{x}$ can be written as $Q \equiv \vec{x}^T W \vec{x} + \vec{B} \cdot \vec{x} + C$; or equivalently, since W is positive definite and therefore invertible and the constant factor C does not affect the solution, as $Q = (\vec{x} - \vec{x}_0)^T W (\vec{x} - \vec{x}_0)$ where $\vec{x}_0$ is some point in the parameter space. While $\vec{x} = \vec{x}_0$ may

---

[5]An example of incommensurable degrees of freedom is the simultaneous calibration of the various parameters of the Heston model.

[6]By convention, the problem is cast as constrained minimization rather than maximization; thus we should call this a "figure of demerit."

not satisfy the equality constraints, it is assuredly the minimizer of Q. A positive definite matrix, like W, defines a metric on the parameter space; thus Q is the squared distance (under this metric) from the ideal solution $\vec{x}_0$. In short, quadratic programming (QP) is a search for the solution to an underdetermined problem which is *closest to some initial point*.

The constraint function f will obviously be nonlinear in any problem of interest: even in yield curve fitting, a swap rate is not linear in any forward rate. However, a large and relevant class of problems is nearly linear, and can be well approximated by linearizing in the neighborhood of a solution. Thus the solution to the QP problem, using a linearized approximation $\vec{f} \simeq \vec{f}(\vec{x}_0) + J(\vec{x} - \vec{x}_0)$, suggests an improved parameter point

$$\vec{x}_1 \equiv \vec{x}_0 - W^{-1}J^T(JW^{-1}J^T)^{-1}\vec{f}(\vec{x}_0)$$

which, of all solutions to the linearized constraints, is closest to $\vec{x}_0$. The step from $\vec{x}_0$ to $\vec{x}_1$ is the *QP step*, which has the same role in underdetermined search that the Newton step has in the more familiar fully-specified search. A sequence of QP steps leads us to a solution which, while not exactly the closest to $\vec{x}_0$, is a very good working approximation.[7]

7.4.1 *Function and Jacobian*

The Jacobian computation is typically the most time-consuming part of this search, especially when we must resort to finite differencing. There are two ways we might optimize the computation of J: by having a fully analytic result, or by having a rapidly computable approximate objective function $\tilde{f} \simeq \vec{f}$ and computing its Jacobian by finite differencing. Obviously the first is preferable, but not always feasible.

Whenever possible, we use Broyden's update of the Jacobian rather than compute it anew:

$$J \simeq \tilde{J} \equiv J + \frac{(\delta\vec{f} - J\delta\vec{x}) \otimes \delta\vec{x}}{||\delta x||^2}$$

where $\delta\vec{x} \equiv \vec{x}_1 - \vec{x}_0$ and $\delta\vec{f} \equiv \vec{f}(\vec{x}_1) - \vec{f}(\vec{x}_0)$ — exactly as in the fully-specified case.[8] We must monitor the approximate Jacobians to ensure that they are not too inaccurate, by comparing the expected progress toward the root with that actually obtained.

---

[7]This solution, once found, can in principle be "polished" to move closer to $\vec{x}_0$ without altering $\vec{f}$. In practice this has little value.

[8]It is tempting to try Broyden's bad update here, since it has the property of seeking minimal change to the solution. I have not attempted this.

This dictates the interface of the class we will use to represent a function to the rootfinder.[9] We will enclose this within namespace Underdetermined to keep class names brief.

```
───────────── Underdetermined.h ─────────────
class Function_
{
   virtual double BumpSize() const;
   virtual void FFast(const Vector_<>& x, Vector_<>* f) const
5      { *f = F(x); }
public:
   virtual ~Function_();
   virtual Vector_<> F(const Vector_<>& x) const = 0;
   virtual void Gradient
10      (const Vector_<>& x,
         const Vector_<>& f,
         Matrix_<>* j)
      const;
};
```

The default implementation of Gradient uses finite differencing of FFast — so we may override either of those functions, but generally not both at once. If neither is overridden, the method reduces to finite differencing of f. The base class supplies a default BumpSize for any necessary finite differencing.

Usually the weight matrix W will be quite sparse, so the dominant storage requirement is $O(Mn)$ for the Jacobian. If we need to solve a truly large problem – e.g., time-dependent path reweighting with $M \gtrsim 10^5$ – then keeping the entire Jacobian in an unstructured dense matrix is excessively memory-intensive. Thus we need to abstract the Jacobian:

```
───────────── Underdetermined.h ─────────────
class Jacobian_
{
public:
   virtual ~Jacobian_();

5   virtual int Rows() const = 0;
   virtual int Columns() const = 0;

   virtual void DivideRows(const Vector_<>& tol) = 0;
10   virtual Vector_<> MultiplyRight(const Vector_<>& t)
      const = 0;
   virtual void QForm
      (const Sparse::Square_& w,
         SquareMatrix_<>* form) const = 0;
15   virtual void SecantUpdate
```

───────────────────────────────

[9]Yes, I just criticized this approach for one-dimensional searches. However, the same gains in simplicity and transparency are not available here.

```
     (const Vector_<>& dx, const Vector_<>& df) = 0;
};
```

In many cases the Jacobian can be implemented with $O(M)$ storage, like W, so the total memory requirement is $O(M + n^2)$. Such a Jacobian obviously must be computed within the Function_, so we add to Function_ an additional member:

─────────── Underdetermined.h ───────────
```
   virtual Jacobian_* Gradient
     (const Vector_<>& x,
      const Vector_<>& f)
   const {return nullptr;}
```

The search engine will call this function, and then if it returns nullptr will call the dense-matrix implementation of Gradient.

7.4.2   Weights and Smoothing

We have talked about finding the closest solution to some initial guess, but have begged the question of what "closest" means except that it is defined in the metric induced by W. It turns out that control of W is the tool we need to obtain *smooth* solutions.

For concreteness, consider fitting an ordered sequence of terms, like forward rates along a yield curve. If we take

$$s^T W s = \sum_i s_i^2 + \lambda \sum_i (s_i - s_{i-1})^2,$$

then W penalizes any change (the first sum) and additionally penalizes nonsmooth changes (the second term). This forms a tridiagonal matrix W with a smoothing parameter which the user can supply. As $\lambda \to 0$, we are simply asking for the smallest aggregate change in the L_2 measure; as $\lambda \to \infty$, the smoothest (smallest aggregate change to differences). The well-known trade-off between smoothness and locality is here made quantitative and explicit.

Piecewise constant forward curves – for rates, spreads, vols, and so on – are an important special case, but we cannot assume that the pieces will be of equal size. In general $\vec{x}$ describes a curve $g(t)$ such that $g(t) = x_i$ in the interval $[T_{i-1}, T_i)$.[10] Then the diagonal part of W corresponds to $\int |g|^2$, while the off-diagonal part corresponds to $\int |g'|^2$. While the latter

─────────────────────

[10]In a well-parametrized models only integrals of g will be observable, so the exact bounding of the interval need not concern us.

is not defined for discontinuous functions, we can achieve the desired effect by computing it for a linear interpolant connecting the midpoints of the intervals; we obtain

$$s^T W s = \sum_i (T_i - T_{t-1}) s_i^2 + \lambda \sum_i \frac{(s_i - s_{i-1})^2}{T_i - T_{i-2}}$$

where λ now has units of t^2. Thus the user-input parameter is a *smoothing timescale* $\tau_s \equiv \sqrt{\lambda}$.

This is implemented (again, in namespace Underdetermined) as:

```
────────────────── UnderdeterminedUtils.cpp ──────────────────
Sparse::Tridiagonal_* WeightsPWC
    (const Vector_<DateTime_>& knots, double tau_s)
{
    unique_ptr<Sparse::Tridiagonal_> retval
        (new Sparse::Tridiagonal_(knots.size()));
    SelfCouplePWC(retval.get(), knots, tau_s, 0);
    return retval.release();
}
```

This forwards to a more general coupling function, supplying an additional argument which is an offset within the matrix being formed. Thus we are prepared for the task of simultaneously fitting two or more curves; for example, in calibrating the two vol curves of a Vasicek model (Sec. 13.1).

The coupling functions rely heavily on a utility function in namespace Sparse,

```
────────────────── SparseUtils.h ──────────────────
inline void AddCoupling
    (Square_* dst, int i, int j, double amount)
{
    dst->Add(i, i, amount);
    dst->Add(i, j, -amount);
    dst->Add(j, i, -amount);
    dst->Add(j, j, amount);
}
```

This reflects how we actually form sparse matrices, and saves us the tedious job of ensuring their symmetry.

7.4.3 Monitoring Progress

To make sure that we are progressing towards a root, we extend the technique of *backtracking linesearch* used in multi-dimensional root search. Given two points, say $\vec{x}_0$ and $\vec{x}_1$, and their corresponding function values, we define $\tilde{f}(k) \equiv k\vec{f}(\vec{x}_0) + (1-k)\vec{f}(\vec{x}_1)$. Then we compute the *overshoot fraction* $k_{\min}$, the value of k which minimizes $|\tilde{f}|$.

We may choose to backtrack, using some *backtracking fraction* $\bar{k}$: that is, replace x_1 with $(1-\bar{k})x_1 + \bar{k}x_0$, evaluate $\vec{f}$ again at this new point, and restart the backtracking process.

At this point we descend from the realm of principle to that of engineering. We wish to backtrack when our step is substantially too large; but since backtracking necessitates an additional function evaluation, we allow some overshoot. Conversely, if $k_{\min} > \frac{1}{2}$ then $\vec{f}(\vec{x}_1)$ is actually further from the solution than $\vec{f}(\vec{x}_0)$. In this case we will should discard our Jacobian and compute a new one; but (especially if the Jacobian is already new) we will first take the best step we can find along the linesearch.

We will backtrack by less than the overshoot fraction in many cases, to make $\bar{k}$ a continuous function of $k_{\min}$. Such continuity, which increases the likelihood that small perturbations will have correspondingly small effects, is important for the stability of calibrations (so that, for example, repeated updates of live pricing requests in changing market conditions will be trustworthy). This is one of many cases where we are willing to sacrifice some accuracy for stability. We can simply take $\bar{k} = 2(k_{\min} - k_{tol})$ when $k_{tol} < k_{\min} < 2k_{tol}$, for some control parameter $k_{tol} < \frac{1}{4}$.

The rootfind succeeds when $|f_i| < \epsilon_i$ for all i, for a user-supplied tolerance $\vec{\epsilon}$. It fails if we exhaust pre-determined limits on the allowed number of function evaluations or gradient evaluations, or if we cannot decrease $|f|^2$ even after frantically backtracking – *i.e.*, if the Jacobian freshly evaluated by `Gradient` does not point in a descent direction.

Thus the search routine has many control parameters, which we encapsulate in a structure (again in `namespace Underdetermined`) with sensible values provided by a default constructor.[11] In fact, we do not write this class, but describe it using mark-up:

―――――――――― *Underdetermined.h* ――――――――――

```
/*IF-----------------------------------------------------
settings UnderdeterminedControls
    controls for underdetermined search
&members
maxEvaluations is integer
    Give up after this many point evaluations
maxRestarts is integer
    Give up after this many gradient calculations
maxBacktrackTries is integer default 5
```

―――――――――――――――――――――――――――――――――――――

[11] We can also change the search engine to extend too-short steps (those with $k_{\min} < 0$); this will of course be controlled by still more parameters.

```
10      Iteration limit during backtracking linesearch
    restartTolerance is number default 0.4
        Restart when k_min is above this limit
    ...
    &conditions
15  maxEvaluations_ > 0
    maxRestarts_ > 0
    restartTolerance_ >= 0.0 && restartTolerance_ <= 1.0
    -IF-----------------------------------------------------*/
```

This defines a `struct` which can be formed from a user-input `Dictionary_`, or formed internally and altered as needed. Our interface parser will create a corresponding `struct`, constructible from and convertible to a `Dictionary_`; and will invoke the required translation code at the public interface.

The control parameter `restartTolerance_` (or k_r) is the value of k_{min} above which we force a "restart", a new call to `Gradient` to refresh the Jacobian. We may introduce a second parameter, `restartToleranceApproxJ` or k_{rs}, to be used when the Jacobian is already approximate (having been updated with Broyden's method). These control parameters are related by $\frac{1}{2} \geq k_r \geq k_{rs} \geq 2k_{tol}$; *i.e.*, we are more willing to restart when we have not just restarted, and we attempt full backtracking ($\bar{k} = k_{min}$) before agreeing to a restart.

This almost completes the interface of our search function.

```
────────── Underdetermined.h ──────────
Vector_<> Find
    (const Function_& func,
    const Vector_<>& guess,
    const Vector_<>& tol,
5    const Sparse::Square_& w,
    const Controls_& controls,
    Matrix_<>* eff_j_inv = nullptr);
```

The last argument, which asks for the final value of $W^{-1}J^T(JW^{-1}J^T)^{-1}$, is used to prepare for repeated calibrations – *e.g.*, when computing vega risk by bumping each calibration instrument price.

7.5 Quadrature

Quadrature, or numerical integration, is the mirror image of root search; as with root search, we write an integrator interface so as to avoid having to wrap the integrand in its own class. Two additional complications arise:

the integrand may be vector-valued, and the integration may be adaptive. We also might wish to reuse an integrator (whereas a rootfinder is always used and then thrown away).

```
_____ Quadrature.h _____
  template<class T_> class Quad1D_ : public Quad1DBase_
  {
  public:
      virtual double GetX() = 0;
5     virtual void PutY(const T_& y) = 0;
      virtual bool IsComplete() const = 0;
      virtual T_ Result() const = 0;
      virtual void Restart() = 0;
  };
```

Clearly `Result` should be called only if `IsComplete` returns `true`; we will check this with an assertion inside implementations of `Result`. The `Restart` function restores the object to its newly-constructed state for reuse. The base class `Quad1DBase_` exists only to provide a virtual destructor.

7.5.1 *Gaussian Quadrature*

A fixed integrator (*e.g.*, Gauss-Hermite) sets the abcissas without any feedback from the function; it is just a weighted sum of evaluations. We represent this in a derived class:

```
_____ Quadrature.h _____
   template<class T_> class Quad1DFixed_ : public Quad1D_<T_>
   {
       size_t i_;    // because STL sizes are size_t
       T_ sum_, initial_;
5  protected:
       Vector_<> x_, w_;
       Quad1DFixed_(int size, const T_& initial)
           : x_(size), w_(size), i_(0),
             initial_(initial), sum_(initial)
10     {  }
   public:
       double GetX() { assert(!IsComplete()); return x_[i_]; }
       void PutY(const T_& y)
       {
15         assert(!IsComplete());
           Quadrature::Increment(&sum_, y, w_[i_++]);
       }
       bool IsComplete() const { return i_ == x_.size(); }
       T_ Result() const { assert(IsComplete()); return sum_; }
20     void Restart() { i_ = 0; sum_ = initial_; }
       // allow query
```

```
    const Vector_<>& Abcissa() const { return x_; }
    const Vector_<>& Weight() const { return w_; }
};
```

A specific integrator, such as Gauss-Legendre, is now implemented as a subclass of this function whose constructor calls a (non-template) function to populate the `abcissa_` and `weight_` vectors. Thus the complexity of Gaussian quadrature is encapsulated in these few functions, while the scaffolding of integration over predetermined abcissae is collected in the template classes we have just displayed.

This implementation is supported by a function to encapsulate `s += w * y`:

```
───────────────── Quadrature.h ─────────────────
namespace Quadrature
{
    template<class T_> inline void Increment
        (T_* dst, const T_& inc, double w)
    {
        T_ z(inc); z *= w; *dst += z;
    }
    template<> inline void Increment    // specialization
        (Vector_<>* dst, const Vector_<>& inc, double w)
    {
        Transform(dst, inc, LinearIncrement(w));
    }
}
```

The template specialization avoids an unnecessary vector copy when the integrand is a `Vector_<>`.

It is worth saying a few words about Gauss-Hermite integration in particular. The use of this in practice is for computing the expectation of a function of a normal deviate; but the conventions of Gauss-Hermite (with weight $\exp\{-x^2\}$) are off by constant factors in both x and w from those of the normal distribution (with $\phi(x) = \exp\{-x^2/2\}/\sqrt{2\pi}$). Rather than make repeated *ad hoc* corrections for this, it is better to have an integrator class take care of it once and for all. Similarly, we will wrap Gauss-Laguerre integration in `class GammaExpectation_`, which comports with its most common use.

```
───────────────── QuadratureGaussian.h ─────────────────
template<class T_ = double> class NormalExpectation_
        : public Quad1DFixed_<T_>
{
public:
    NormalExpectation_(int n, const T_& initial = 0.0)
```

```
         : Quad1DFixed_<T_>(n, initial)
     {
         Quadrature::NCDFGaussHermiteWeights(&x_, &w_);
     }
10   };
```

We supply 0.0, not a more general default, because there is no good default value for vector integrands: we must start with a vector of the appropriate size. Our code gives a compilation error if we attempt

```
NormalExpectation_<Vector_<>> integrator(10);
```

but allows the obvious default for scalar integrands.

7.5.2 *Adaptive Quadrature*

Adaptive integrators, such as Romberg's, are intrinsically unstable: for two near-identical inputs, where we would want to see near-identical results (*e.g.*, when computing a parameter sensitivity by finite differencing), the signal can be swamped by a change in the integrator's strategy. The signature of this instability is infrequent and seemingly random blowouts in computed risk figures. This can be avoided with sufficient machinery (see Secs. 5.5, 10.12, and 14.4); but for most applications we will favor fixed integrators for their greater speed and stability, despite the lack of guaranteed convergence.

7.6 Distributions

Another common building block is a one-dimensional (likely risk-neutral) probability distribution. The lognormal distribution is the canonical example. We begin with an abstract base class, whose specification reveals our intentions for its use:

```
──────────────── Distribution.h ────────────────
class Distribution_
{
public:
    virtual ~Distribution_();
5   virtual double Forward() const = 0;
    virtual double OptionPrice
        (double strike, const OptionType_& type)
    const = 0;
    // support calibration of whatever vol-like parameter
```

```
10    virtual double& Vol() = 0;    // whatever it means
      virtual const double& Vol() const = 0;
      virtual double VolVega
          (double strike, const OptionType_& type)
      const = 0;
15    // support hedge computation
      virtual Vector_<String_> ParameterNames() const = 0;
      virtual map<String_, double> ParameterDerivatives
          (double strike,
          const OptionType_& type,
20        const Vector_<String_>& to_report)
      const = 0;
};
```

We assume that every distribution has a unique vol-like parameter: lognormal vol, normal vol, the SABR α, and so on. The oddly named `VolVega` is a *proportional vega*, the product of vol and vega (or the derivative of price with respect to log of vol). This turns out to be much more useful than vega alone for calibration; it is very similar across different distributions with different meanings of "vol" (*e.g.*, normal and lognormal distributions). Also, it has no units of time, so it is the same for annualized or deannualized vols; this property is useful when implementing stochastic-time distributions (such as gamma variance) which must themselves report some form of vega.

The last two functions will be used to support "market models" for swaptions, equities or FX – these are distribution-based pricing models with no specification of the dynamical process from which that distribution arises, and thus are valid only for simple European options.[12] Armed with this functionality, we can build the framework of such a model without committing to any particular parametrization of the distribution; see Sec. 14.6.

7.6.1 *Implied Vol*

One recurrent problem is the need to compute reliable implied vols, based on a user-specified distribution, for deep in- or out-of-the-money options which have extremely little value; the option price may be indistinguishable from the intrinsic value. A straightforward implied vol computation will thus produce numbers which are correct, in the sense that they reproduce the option price to good accuracy, but are less than intuitive. We can

---

[12]Jamshidian *et al.* have attempted to back out curve dynamics from swap market models, but I am unaware of any practical tools to emerge from this.

choose how to spend our time: in explaining to users why this does not
matter, or in implementing a more robust implied vol.

Given a distribution (rather than just a single option price), we can
accomplish the latter goal by stepping outward from the forward to the
actual strike, using the implied volatility from the previous iteration (at
first, the at-the-money vol) as the initial guess each time. To accomplish
this, we write

```
─────────────── DistributionUtils.cpp ───────────────
double Distribution::BlackIV
    (const Distribution_& model,
     double strike,
     double guess,
5    int n_steps)
{
     const double f = model.Forward();
     const OptionType_ type = strike > f
             ? OptionType_::Value_::CALL
10           : OptionType_::Value_::PUT;

     if (n_steps > 1)
     {
         const double fMid = strike > f
15               ? strike * pow(f / strike, 1.0 / n_steps)
                 : strike + (f - strike) / n_steps;
         guess = BlackIV(model, fMid, guess, n_steps - 1);
     }
     return BlackIV
20        (f, strike, type, model.OptionPrice(strike,type), guess);
}
```

where the last call is to the usual form of `BlackIV`. This is a non-tail
recursion, and it is not worth the effort of unwinding it into an iteration;
usually half a dozen steps gives an answer of very good quality.

7.7 Baskets

Basket options are an important class of equity derivatives; they are also
a crucial tool for swaption pricing in Libor-based interest rate models
(Goldman-Pugachevsky, BGM or "string" models) where a swap is approximated as a basket of FRAs (forward rate agreements) or futures. Both
problem descriptions are similar: we have a collection (the "basket") of
assets with similar dynamics, and an option on a weighted sum of their
values, so we must approximate $E\left[(B - K)^+\right]$ where $B \equiv \sum_i w_i S_i$.

Clearly the weights are not central to the problem, since we can absorb the scale factor into S_i. Once the dynamics of the S_i are specified, we can take one of two approaches: *whole-basket* methods which approximate the distribution of the total basket value, or *effective-strike* methods which use the strike K to extract relevant statistics for each S_i. In our experience, effective-strike methods are faster and better behaved in the extreme tails of the distribution, though they are trickier to implement.

7.7.1 *Whole-Basket Moment Matching*

If the S_i are multivariate normal, then B is also normal and the problem is trivial. If they are multivariate lognormal, then B is not lognormal, but we can approximate it by a shifted lognormal. For this purpose we need the first three moments of B. These can be accumulated in a simple loop:

```cpp
                        ───── Basket.cpp ─────
  *m1 = *m2 = *m3 = 0.0;
  double m22 = 0.0;
  double m32 = 0.0, m33 = 0.0;
  for (int ii = 0; ii < n; ++ii)
5 {
      if (IsZero(wf[ii]))
          continue;
      const double& si = wf[ii];
      *m1 += si;
10    *m2 += Square(si) * ec(ii, ii);
      *m3 += Cube(si * ec(ii, ii));
      for (int jj = 0; jj < ii; ++jj)
      {
          if (IsZero(wf[jj]))
15            continue;
          const double& sj = wf[jj];
          m22 += si * sj * ec(jj, ii);
          m32 += si * sj * Square(ec(jj, ii))
                    * (si * ec(ii, ii) + sj * ec(jj, jj));
20        for (int kk = jj + 1; kk < ii; ++kk)
              m33 += si * sj * wf[kk]
                        * ec(jj, ii) * ec(kk, ii) * ec(jj, kk);
      }
  }
25 *m2 += 2.0 * m22;
  *m3 += 3.0 * m32 + 6.0 * m33;
```

Here `ec` is the matrix of exponentiated covariances:

$$\mathtt{ec(i, j)} = \exp\Big\{ \mathrm{Cov}[\log S_i, \log S_j] \Big\}.$$

Creating this matrix saves $O(n^3)$ calls to `exp`, so it is a very worthwhile optimization even if we must allocate the matrix. The choice of the loop conditions for j and k (so that $j < k < i$, rather than the more usual $k < j < i$) is due to the routine's importance for swaption pricing: we can have several zero weights at the front (before expiry) and the back (after swap maturity) of the basket, and we can omit these terms without having to put an extra test in the innermost loop.

In practice this clean inner loop will be obscured by the need to accumulate optional delta and vega outputs. We will translate the moments to a lognormal part (forward and vol) and a shift, and use these in pricing; sensitivity to these parameters can then be mapped back to sensitivity to the moments themselves, and thence to the input parameters, by repeated application of the chain rule. Thus the relevant intermediate quantity is the sensitivity of each moment to each of the forward prices $E[S_i]$ or covariances; we will accumulate these within the loop and combine them at the end, just as we have done for m_2 and m_3.

Fitting a shifted lognormal distribution to the computed moments requires the solution of a cubic equation. Let $y \equiv \exp\{\sigma^2\}$ where σ is the volatility of the lognormal part; then

$$(y-1)(y+2)^2 = \frac{\left[\frac{m_3 - m_1^3}{m_2 - m_1^2} - 3m_1\right]^2}{m_2 - m_1^2}.$$

If we are computing sensitivities, the root of the cubic appears again in the chain-rule application; thus it must be provided as an optional output.

```
------------------------- Basket.cpp -------------------------
void XThreeMomentFit
    (double m1, double m2, double m3,
     double* ln_part, double* shift, double* vol,
     double* root);
```

7.7.2   Taylor Expansion of Projected Vols

If we define

$$B \equiv \sum w_i S_i, \qquad Q_i \equiv \sum_j \sigma_{ij} w_j S_j, \quad \text{and} \quad V \equiv Q \cdot S = (wS)^T \sigma(wS),$$

then we can compute the elasticity of the basket price B:

$$\beta = \frac{1}{2} \frac{dV \wedge dB}{dB \wedge dB} \left(\frac{V}{B}\right)^{-1} = \frac{(\sum_j Q_j^2 w_j S_j)(\sum_j w_j S_j)}{(\sum_j Q_j w_j S_j)^2}.$$

We will always have $\beta \geq 1$ by Cauchy's inequality. Thus we can approximate the basket as a shifted lognormal distribution, where the shift is $\Delta \equiv B(1 - \beta^{-1})$. The volatility of the lognormal can be set to $\sqrt{V}/(B - \Delta)$, which is correct at time 0, or by matching the second moment of the total basket price. This computation takes only $O(n^2)$ time, and is competitive with whole-basket methods in accuracy.

7.7.3 *Midpoint Variance*

A better method for matching the total basket variance is to use the strike; or, more specifically, the characteristics of the mean path from spot to strike. We are interested in the basket's volatility when $B \simeq (E[B] + K)/2$; analysis of this midpoint captures most of the benefits of path integration.

Define the relative strike $M = K - E[B]$, and for each i compute the "strikelet"

$$\tilde{k}_i \equiv E\left[S_i | B = K\right] = E[S_i] + \frac{MQ_i}{w_i \sum_j Q_j} + O(M^2).$$

We can convert each lognormal volatility to a normal volatility by taking the instantaneous conversion at the midpoint, $\sigma_i^{(N)} = \sigma_i^{(ln)}(E[S_i] + \tilde{k}_i)/2$; or by inferring a normal vol from the value of the option struck at $\tilde{k}$. The latter approach is more computationally intensive, but more robust for highly inhomogenous baskets.

7.8 Random and Quasi-Random Numbers

7.8.1 *Random Deviate Streams*

A random[13] generator should encapsulate both the production of uniform deviates and their transformation into normal deviates. For most dynamical models, we will draw batches of normal deviates; however, uniform deviates are more convenient for simulation of discrete events (jumps or default). We will support this with the following interface:

—————————————— *Random.h* ——————————————
```
class Random_
{
public:
    virtual ~Random_();
```

---

[13]These are deterministic and therefore, strictly speaking, only "pseudo-random." We nonetheless prefer the briefer nomenclature.

```
 5   virtual double NextUniform() = 0;
     virtual void FillUniform(Vector_<>* deviates) = 0;
     virtual void FillNormal(Vector_<>* deviates) = 0;
     virtual Random_* Branch(int i_child = 0) const = 0;
   };
```

The first three functions draw random numbers, changing the class state in the process. **Branch** creates an independent random stream; see Sec. 10.11.

A sensitivity is computed from the difference of the price computed with the input model (the *base run*) and the same price computed with a slightly different *bumped model* (the *bumped run*). We want to exclude any source of variation in price except for the difference in models; thus, for instance, we want to use the same deviates in the same order.

We might think to store all the random deviates used in the base run, and simply apply them again during the bumped run. This approach has two related weaknesses. First, the extra memory requirement reduces the maximum number of paths we can run on a given machine, which for long-running trades in very complex models is already an unpleasant constraint. Second, the additional memory requirement increases the overhead of the computation, replacing productive CPU usage with cache faults, so it is probably faster to regenerate the deviates than to save them.

Whether random deviates are stored or re-created, the pattern in which they are used – the mapping from the random stream to financial observables – must be unchanged. We accomplish this by using the same number of deviates at each step on each path, so that the state of the random stream will be the same at the corresponding point in both runs.

7.8.2   *Generator Implementation*

Though **FillUniform** is declared pure virtual, we will supply the obvious implementation:

```
                         ———— Random.cpp ————
   void Random_::FillUniform(Vector_<>* devs)
   {
      for (auto ud : *devs)
         ud = NextUniform();
 5 }
```

This function can be called explicitly by a derived class, using the syntax `Random_::FillUniform(deviates)`. It is not supplied as a full-fledged default because of its inefficiency: it requires a virtual function call for each deviate. A production generator should avoid this.

There are several good choices for generation of uniform deviates. Our favorite methods are

(1) The "Mersenne twister" of Matsumoto and Nishimura, our first choice.
(2) Numerical Recipes Ran2, a reliable standard.
(3) A pair of generators like Knuth's IRN55, with the second shuffling the outputs of the first.

All these are readily available online.

The option to switch generators at runtime seems to have minimal value, so we prefer to settle on one of these generators and use it for all pricing. Thus we create random generators through the interface:

```
                        Random.h
namespace Random
{
    Random_* New(int seed);
}
```

There is no need for seed to be negative, since we have abandoned the (very restrictive) trick of using it to smuggle the random generator's state. Acclimation to a world of positive seed values is quick and painless.

7.8.3 *Transforms*

Another question is how to effect the transformation of uniform to normal deviates. Again, there are several plausible candidates:

(1) Directly through the inverse cumulative normal density.
(2) The polar method of Box and Mueller, most widely used.
(3) The odd-even method of Von Neumann and Forsyth.
(4) The rectangle-wedge-tail method of Marsaglia, the fastest available.

Knuth (in *Art of Computer Programming*, volume 2) gives an implementation of Marsaglia's method but leaves computation of its many constant parameters as an exercise for the reader. In addition, Knuth's routine partitions the distribution into 32 segments; by using a finer partition, we increase the size of the tables of constants but decrease the frequency of calls to the math library. For example, with partition into 64 segments we have the static data (in **namespace RWT**):

```
                        Random.cpp
static const double D[31] = { 0.0, 0.0, 0.0, 0.0, 0.0, 0.0, 0.0,
    0.0, 0.0, 0.0, 0.0, 0.0, 0.0, 0.0, 0.0, 0.0,
```

```
      0.5050335007, 0.7729568318, 0.8764243173, 0.9392112429,
      0.9860868156, 0.9951545013, 0.9867480142, 0.9792113586,
5     0.9722739162, 0.9657523400, 0.9595309729, 0.9535340961,
      0.9477102649, 0.9420234020, 0.9364475249 };
    static const double E[31] = { 0.0, 0.0, 0.0, 0.0, 0.0, 0.0, 0.0,
      0.0, 0.0, 0.0, 0.0, 0.0, 0.0, 0.0, 0.0, 0.0,
      25.0, 12.5, 8.3333333333, 6.25, 5.0, 4.0637731069,
10    3.3677961409, 2.8582959135, 2.4694553648, 2.1631696640,
      1.9158499112, 1.7121118654, 1.5414940825, 1.3966346593,
      1.2722024279 };
    static const double P[32] = { 0.0, 0.8487410443, 0.9699899979,
      0.8550231215, 0.9942754672, 0.9951625205, 0.9327422986,
15    0.9233994671, 0.7273661575, 1.0, 0.6910843714, 0.4540747884,
      0.2866499878, 0.1738620062, 0.1013177803, 0.0567276597,
      0.0672735098, 0.1605108070, 0.2355403454, 0.2854029087,
      0.3075794736, 0.3038922909, 0.2795217494, 0.2414883115,
      0.1970555059, 0.1524486512, 0.1121116518, 0.0785256947,
20    0.0524616474, 0.0334682128, 0.0204066391, 0.0863979023 };
    static const double Q[16] = { 0.0, 0.2356431344, 0.2061876931,
      0.2339118030, 0.2011514983, 0.2009721989, 0.2144214970,
      0.2165909849, 0.2749646762, 0.2, 0.2894002647, 0.4404560771,
      0.6977150132, 1.1503375831, 1.9739871865, 3.5256169769 };
25  static const double S[17] = { 0.0, 0.0, 0.2, 0.4, 0.6, 0.8, 1.0,
      1.2, 1.4, 1.6, 1.8, 2.0, 2.2, 2.4, 2.6, 2.8, 3.0 };
    static const double Y[32] = { 0.0, -0.922235758, -5.864444728,
      -0.579530323, -33.13734925, -39.54384419, -2.573637156,
      -1.610947892, 0.666415357, DA::INFINITY, 0.352574032,
30    -0.166350547, 0.919632724, 0.357909694, -0.022548077,
      0.187972157, 0.585574869, 0.961759887, -0.061622701,
      0.120122007, 1.311158187, 0.312688141, 1.12240661, 0.536325751,
      0.75091678, 0.564026097, 0.174746453, 0.382956509, -0.01107325,
      0.393074576, 0.195833651, 0.781086317 };
35  static const double Z[32] = { 0.2, 1.3222357584, 6.6644447279,
      1.3795303233, 34.9373492509, 41.3438441882, 2.9736371558,
      2.6109478918, 0.7335846430, DA::INFINITY, 0.6474259683,
      0.3663505472, 0.2803672763, 0.2420903063, 0.2225480772,
      0.2120278433, 0.2144251312, 0.2382401128, 0.2616227015,
40    0.2798779934, 0.2888418127, 0.2873118590, 0.2775933900,
      0.2636742492, 0.2490832199, 0.2359739033, 0.2252535472,
      0.2170434909, 0.2110732504, 0.2069254241, 0.2041663490,
      0.2189136830 };
```

This supports the algorithm, which closely follows Knuth:

```
                        Random.cpp
template<class SRC_> FORCE_INLINE void Fill
    (SRC_* src,
     Vector_<>::const_iterator dst_begin,
     Vector_<>::const_iterator dst_end)
5  {
      for (auto pn = dst_begin; pn != dst_end; ++pn)
```

```
 {
        double f = 64.0 * src->NextUniform();
        double sign = (f > 32.0 ? 1.0 : -1.0);
10      int j = int(f);
        f -= j;
        j &= 31;
        // f is uniform in [0, 1), j in {0..31}
        if (f >= P[j])    // 60.55%
15          *pn = Y[j] + f * Z[j];
        else if (j < 16)      // 31.28%
            *pn = S[j] + f * Q[j];
        else if (j < 31)        // 7.90%
        {
20          double u, v;
            do      // loop c. 1.056 times
            {
                u = src->NextUniform();
                v = src->NextUniform();
25              if (u > v)
                    swap(u, v);
                *pn = S[j - 15] + 0.2 * u;
                if (v <= D[j]) // In triangle
                    break;
30          }   // full rejection test:  1.00%
            while ((exp(0.5 * (Square(S[j - 14]) - Square(*pn)))
                    - 1.0) * E[j] + u < v);
        }
        else    // "Supertail" case; 0.27%
35      {
            do   // loop c. 1.094 times
            {
                const double u = src->NextUniform();
                *pn = sqrt(9.0 - 2.0 * log(u));
40          } while (*pn * src->NextUniform() >= 3.0);
        }
        *pn *= sign;
    }
}
```

The `inline` directive may not be strong enough; many compilers will judge the code too complex for inlining. Thus we use `FORCE_INLINE`, a macro which will be resolved as a platform-dependent directive. This routine uses on average about 1.17 calls to `NextUniform`, and calls a math library function (`exp`, `sqrt` or `log`) once for every 60 normal deviates.

If the Marsaglia algorithm is deemed too complex, the strongest alternative is the direct method, due to its simplicity of implementation.

7.8.4   *Low-Discrepancy Sequences*

The convergence of Monte Carlo methods can in many cases be greatly improved by using numbers which, rather than emulate randomness, approach a uniform distribution over the domain of integration (which for this purpose is mapped onto the D-dimensional unit hypercube, $[0, 1]^D$). Several families of such sequences are known. A few are not useful for one reason or another:

- *Halton sequences* depend on a different prime base for each dimension; if the dimension is significant, then the prime bases become large and successive paths slowly "scan" along a few low-dimensional manifolds within the unit cube.
- *Faure sequences* depend on a prime base $P \geq D$. The coverage after P^{P-1} paths is excellent; then the next P^{P-1} paths nearly duplicate the previous, forming a smaller-scale cover which is not complete until we reach P^{P^2-1} paths. This discourages small values of P, but for large P we again see the "scanning" behavior. Faure sequences are also not well adapted to binary computation.

The most useful low-discrepancy sequences are those of Sobol and of Niederreiter. They seem to perform equally well in practice; here we will focus on Sobol sequences, whose implementation is somewhat simpler.

We must always bear in mind that Sobol (or other) sequences are an intrinsically unique resource; we must ensure that a given sequence is used only once in a Monte Carlo computation. This is simple enough for simple models, but requires care when we start constructing hybrid models with multiple components, each of which wishes to use random or quasi-random numbers.

We support this resource protection by defining the functionality of a set of sequences:

──────────────────── *QuasiRandom.h* ────────────────────

```
     namespace QuasiRandom
     {
        class SequenceSet_ : noncopyable
        {
5    public:
           virtual ~SequenceSet_();
           virtual int Size() const = 0;
           virtual void Next(Vector_<>* dst) = 0;
           virtual SequenceSet_* TakeAway(int subsize) = 0;
10       };
     }
```

Each call to Next populates dst with numbers in $(0, 1)$, and advances the path index or state of the sequences as necessary. The TakeAway function partitions the sequence set; its input is the number of sequences to place into the returned subset, while the remaining sequences are kept in this. Thus TakeAway(Size()) returns a copy of the existing sequences, emptying this itself in the process. To support hybrid models as described above, we will construct a large SequenceSet_ at the top level, and call TakeAway to extract non-interfering subsets for each submodel.

A Sobol sequence is generated from a set of *direction numbers*, which in turn obey a recursion based on a given primitive polynomial modulo 2. We use one sequence for each polynomial: two sequences based on the same polynomial, even with different initial direction numbers, provide very poor coverage of the unit square.

Once the direction numbers are known, the generating polynomial is no longer needed. Thus our Sobol sequence set does not contain it.

```
                    ───────── Sobol.cpp ─────────
    struct SobolSet_ : QuasiRandom::SequenceSet_
    {
        Matrix_<int> directions_;   // index as [i_bit][i_seq]
        int iPath_;
5       Vector_<int> state_;

        SobolSet_(int i_path) : iPath_(i_path) {}
    public:
        int Size() const override { return state_.size(); }
10      void Next(Vector_<>* dst) override;
        SobolSet_* TakeAway(int subsize) override;
    };
```

We start with a path index i_path $\gg$ 1. The reason is a particular property of Sobol sequences: the first entry in each sequence is $\frac{1}{2}$, which is unusually well centered; then $\frac{1}{4}$ and $\frac{3}{4}$ in either order; and so on. In general the first n paths are lacking in extreme values, and have too little dispersion: the average of $N^{-1}(x_i)^2$ is roughly $1 - \log_2 n/n$. This can distort computed option prices; we avoid the problem by taking our paths – a few thousand in number – from a randomly chosen index among the billion or so possible. To accomplish this, our initializer must be able to "fast forward" to a given path. This turns out not to be difficult:

```
────────────── Sobol.cpp ──────────────
Fill(&seq->state_, 0);
for (int jj = 0, ip = seq->iPath_; ip; ++jj, ip >>= 1)
{
    if ((ip ^ (ip >> 1)) & 1)
    {   // the Gray code of iPath_ has a 1<<jj bit
        Transform(&seq->state_, seq->directions_.Row(jj), XOR);
    }
}
```

This is desirable for another reason: the variation seen by the user as the seed changes should encompass *all* sources of numerical error in the price. If we always start from the first Sobol path regardless of seed, we are hiding such a source of error; thus the average over many runs will converge to a false estimate of the price, while yielding an overoptimistic standard error.

The implementation of Next, following *Numerical Recipes*, is quite efficient:

```
────────────── Sobol.cpp ──────────────
double ScaleTo01(int state)
{
    static const double MUL = 0.5 / (1L << (N_BITS - 1));
    return MUL * state;
}
static const auto XOR = [](int i, int j)->int{ return i ^ j; };

void SobolSet_::Next(Vector_<>* dst)
{
    dst->Resize(Size());    // usually no-op
    ++iPath_;
    assert(iPath_ != 0);    // so next loop can terminate
    int k = 0;
    for (int j = iPath_; !(j & 1); j >>= 1, ++k);
    assert(k < directions_.Rows());
    Transform(&state_, directions_.Row(k), XOR);
    Transform(state_, ScaleTo01, dst);
}
```

Note that we store an int for each sequence's state, but convert to a double in $(0, 1)$ for output. The weird computation of MUL is designed to prevent overflowing the positive range of integers. We use N_BITS == 30; we could grab an extra bit by using unsigned ints, but it has no real value.

For large D, the computation of the first D primitive polynomials can be time-consuming. It is better to compute offline the first few thousand such polynomials and insert them directly into the code.

```
_____ Sobol.cpp _____
static const unsigned int KNOWN_PRIMITIVE[N_KNOWN] =
  {0x0000002, 0x0000003, 0x00000005, 0x000000B, ...
```

We could omit the 1 bit in this representation (adding special handling for the first polynomial $p(x) = x$) — if we were demented enough to foresee the need to handle more than about 50 million sequences.

7.8.5   Spectral and Spining Methods

Sobol (and other low-discrepancy) sequences perform best in low-dimensional problems. As the dimensionality increases, it becomes harder to obtain good convergence even for a simple payout.

This is exacerbated by embedding in our simulation code a naive concept of dimension. If D is simply defined as the number of random deviates consumed in the process of stepping to the trade's maturity, then D will rapidly grow large. But many trades are *directional*, so that the payout of the trade is largely determined by a total (time-integrated) innovation, with little dependence on other details of the path.

We can use directionality to improve convergence, if we can assign Sobol numbers directly to combinations of random deviates which are most important to the trade's payout. This idea, called *reduction of effective dimension*, can be implemented in several ways:

- Via recursive Brownian bridging; this method was pioneered by William Morokoff in the 1990s.
- By a "spining" technique, assigning Sobol deviates to some partial sums of the vector of deviates, developed by the author in 1996.
- By *spectral methods* which assign Sobol deviates to the low-frequency modes of the problem, developed by the author around 2003.

The spectral methods rely on an interesting property of the Fourier transform: namely, the transform of a vector of standard $N(0,1)$ deviates is itself a vector of normal deviates. In my experience, the resulting method is more effective than spining and more generic than Brownian bridge.

For each driver, we will form a path of random deviates, place Sobol deviates at the front, and then Fourier-transform it. There is one wrinkle: since FFT is only available for arrays whose size is a power of 2, we must round up the size before transforming. After transforming, we combine pairs (using $x_i = (y_j + y_{j+1})/\sqrt{2}$) to regain the desired output size.

Extending these techniques to multi-driver models requires some work. Each driver will `TakeAway` some of the available Sobol sequences for its own use, but there is no obvious way to apportion these sequences among the drivers. Equal partition is clearly suboptimal, since the drivers' importance varies both intrinsically to the model (*e.g.*, drivers of higher eigenmodes in an HJM model of the curve are likely less important) and based on trade-model interactions (*e.g.*, interest rate drivers are less important for short-dated equity products).

7.9 PDE Solvers

While our most important models are high-dimensional and require Monte Carlo methods, there are still special cases where finite-difference PDE solvers are valuable. We will have no use for fully explicit "tree" solvers, which are inferior in every application.[14] A full treatment of the science of PDE solvers would require a volume far larger than this one. However, without too much trouble we can build a solver which is unconditionally stable and reasonably accurate. A good general-purpose scheme is described in Lipton's *Mathematical Methods in Foreign Exchange*.

We wish to abstract the PDE solver – a tool to implement a finite-differencing scheme – from the calling routine, which provides the terminal conditions and maybe free boundaries. Passing the terminal condition and coefficient calculators to a "generic" solving function is wrong, for the same reasons discussed in Sec. 7.3: it unnecessarily forces us to bundle the calculators into a class whose very existence is a design flaw. Instead, we encapsulate the scheme as an independent object, derived from an abstract base class in `namespace PDE`:

```
─────────────────────────── PDE.h ───────────────────────────
class Rollback_
{
public:
    virtual ~Rollback_();

    virtual void operator()
        (double dt,          // positive
         const Vector_<CoordinateVector_>& x_points,
         const Vector_<shared_ptr<Cube_>>& old_vals,
         const ScalarCoeff_& discounting,
         const VectorCoeff_& advection,
         const MatrixCoeff_& diffusion,
```

---

[14]See especially our discussion of forward induction, below.

```
           Vector_<shared_ptr<Cube_>>* new_vals)
        const = 0;
};
```

Now let us describe the lower-level interfaces we have just introduced.

7.9.1   *Cube*

A cube is a three-dimensional data structure with dense data; thus it can store the values at each point of a grid of up to three dimensions. This interface choice limits our PDE solver to at most three spatial dimensions; we do not chafe at this restriction, since finite-difference PDE solvers are asymptotically and practically inferior to Monte Carlo beyond this point.

In one dimension, it would seem better to have a `Vector_<>` than a `Cube_`; we minimize this drawback by supporting an efficient `Swap` of the contents of a vector and an $1 \times 1 \times n$ cube. This requires that, at the least, the contents of a one-dimensional "slice" must be stored in a vector; in practice we will store the cube's entire contents in a `Vector_<>`.

We implement `Cube_` as a special case of an N-dimensional array. Unfortunately, our need to efficiently `Swap` with lower-dimensional containers is not supported by the `boost::multi_array` template, so we must roll our own.

```
                          ————— NDArray.h —————
template<class E_> class ArrayN_
{
    Vector_<int> sizes_;
    Vector_<int> strides_;
    Vector_<E_> vals_;  // member ordering used in constructor
public:
    ArrayN_(Vector_<int>& sizes, const E_& fill = 0)
        :
        sizes_(sizes),
        strides_(ArrayN::Strides(sizes)),
        vals_(strides_[0] * sizes[0], fill)
        { }

    const E_& operator[](const Vector_<int>& where) const
        { return vals_[InnerProduct(where, strides_)]; }
    E_& operator[](const Vector_<int>& where)
        { return vals_[InnerProduct(where, strides_)]; }

    inline bool Empty() const { return vals_.empty(); }
    inline void Fill(double val) { vals_.Fill(val); }
    inline void operator*=(double scale) { vals_ *= scale; }
// ...
```

```
      void Swap(Vector_<E_>* other)
      {
25        auto pm = MaxElement(sizes_);
          REQUIRE(*pm == vals_.size(),
              "Can't swap a vector with a multi-dimensional array");
          vals_.Swap(other);
          *pm = vals_.size();
30        strides_ = Strides(sizes_);
      }
// ...
```

We augment this class with functionality to support a more natural interface for arrays like `Cube_`, whose dimension is known at compile time. In this case there is no need to form a `Vector_<int>` every time we wish to represent a location.

```
———————————————— NDArray.h ————————————————
protected:
  struct XLoc_
  {
      int offset_;
5     int sofar_;
      const Vector_<int>& strides_;
      XLoc_(strides_) : offset_(0), sofar_(0) {}
      XLoc_& operator()(int i_x)
      { sofar_ += i_x * strides_[offset_++]; return *this; }
10    int Offset() const
      { assert(offset_ == strides_.size()); return sofar_; }
  };
  XLoc_ Goto() const { return XLoc_(strides_); }
  E_& At(const XLoc_& loc) { return vals_[loc.Offset()]; }
15  const E_& At(const XLoc_& loc) const
      { return vals_[loc.Offset()]; }
};
```

An instance of `ArrayN_::XLoc_` stores (by reference, hence the `X`) the `strides_` of the `ArrayN_` from which it was created. It expects to be given, in sequence, a coordinate value for each dimension of the array – this is checked at runtime by the `assert` within the `Loc` function. Its advantage is that the sequence is of inline function calls, requiring no heap allocation.

This provides enough access that the `Cube_` can implement the 3D interface expected by the PDE engine.

```
———————————————— NDArray.h ————————————————
class Cube_ : public ArrayN_<double>
{
public:
```

```
     Cube_();
 5   Cube_(int size_i, int size_j, int size_k);
     // support lookups without constructing a temporary vector
     const double& operator()(int ii, int jj, int kk) const
     {
         return At(Goto()(ii)(jj)(kk));
10   }
     double& operator()(int ii, int jj, int kk)
     {
         return At(Goto()(ii)(jj)(kk));
     }
15   // allow access to slices (last dimension)
     inline double* SliceBegin(int ii, int jj)
     { return &operator()(ii, jj, 0); }
     inline const double* SliceBegin(int ii, int jj) const
     { return &operator()(ii, jj, 0); }
20   inline const double* SliceEnd(int ii, int jj) const
     { return SliceBegin(ii, jj) + Goto()(0)(1)(0).Offset(); }

     inline int SizeI() const { return Sizes()[0]; }
     inline int SizeJ() const { return Sizes()[1]; }
25   inline int SizeK() const { return Sizes()[2]; }
     void Resize(int size_i, int size_j, int size_k);
};
```

7.9.2   *Coordinate Mapping*

A model communicates coefficients (*e.g.*, the diffusion coefficient $\sigma^2/2$ in a normal model) in its own coordinate space, the same space in which its state variables are specified. We can often improve PDE convergence by using nodes which are not equidistant in this coordinate space; the best way to do so without a great loss of efficiency is to solve the PDE on a regular grid in some transformed coordinate space. This can involve any mapping $\mathcal{R}^n \to \mathcal{R}^n$; in practice we will consider only separable mappings, so we specify a function mapping $\mathcal{R} \to \mathcal{R}$ in each spatial dimension. This means that correlations must be handled by the solver, since they cannot be rotated away in a coordinate transformation.

We also allow the mapping to be time-dependent, but only in a very restricted way: we can rescale all the y_i, at some specified set of times, by some constant factor. The mapping function $x(y)$ is not affected; time-dependent mapping functions do not bring benefits commensurate to their cost in code complexity and execution time. Our more restrictive scheme allows us to "refocus" the grid at short timescales – when the envelope of likely x-values is smaller – at minimal cost.

Our demands on the coordinate mapping are not symmetric between x and y. We need to compute not only $x(y)$ but also its first two derivatives, in order to know the PDE coefficients in y-space; but in the other direction we need only compute $y(x)$.

```
──────────────────────── PDE.h ────────────────────────
class CoordinateMap_ : noncopyable
{
public:
    virtual ~CoordinateMap_();
    virtual double operator()
        (double y, double* dx_dy = nullptr,
         double* d2x_dy2 = nullptr)
        const = 0;
    virtual double Y(double x) const = 0;
};
```

A *coordinate vector* specifies how many points are to be used in a given spatial dimension; their bounds (between which they are evenly distributed) in the solver's coordinate space; and the mapping from solver space (points y_i) to model space (x_i).

```
──────────────────────── PDE.h ────────────────────────
struct CoordinateVector_
{
    double yLow_, yHigh_;
    int n;    // n >= 2 if low != high
    Handle_<CoordinateMap_> yToX_;
    map<DateTime_, double> rescalings_;
};
```

One useful coordinate map is simply $x = \lambda\sinh(y/\lambda)$, which approaches a linear map (equivalent to the identity) as $\lambda \to \infty$. Clearly λ is the necessary member data for this mapping, but we construct it through a factory function which computes λ based on the range of x (maximum absolute value) and of dx/dy (ratio of maximum to minimum).[15]

```
──────────────────────── PDEUtils.cpp ────────────────────────
PDE::CoordinateMap_* PDE::NewSinhMap
    (double x_width,
     double dxdy_range)
{
    assert(IsPositive(x_width) && dxdy_range >= 1.0);
    double sinhMaxY = sqrt(Square(dxdy_range) - 1.0);
    return IsZero(Square(sinhMaxY))
        ? (CoordinateMap_*) new IdentityMap_
```

[15]The cast of the returned IdentityMap_ is needed to allow the two branches of ?: to return the same type.

```
                 : new SinhMap_(x_width / sinhMaxY);
10  }
```

Of course, like any factory function, this lets us keep the class SinhMap_
completely within its source file.

7.9.3 Coefficient Calculators

The coefficient calculators must compute a coefficient which might be
x-dependent, and also state their x-dependence so that the solver can avoid
wasteful recomputation of unchanging coefficients. For a matrix coefficient,
we must state the x-dependence of each element of the matrix. Thus we
write (still in namespace PDE):

```
                          ____ PDE.h ____
   static const size_t MAX_DIMENSIONS = 3;
   class Coeff_
   {
   public:
5      virtual ~Coeff_();
       typedef bitset<MAX_DIMENSIONS> x_dep_t;
   };

   class MatrixCoeff_ : public Coeff_
10 {
   public:
       virtual void Value
          (const Vector_<>& x,
           SquareMatrix_<>* value)
15     const = 0;

       virtual Matrix_<x_dep_t> XDependence() const = 0;
   };

20 class VectorCoeff_ : public Coeff_
   {
   public:
       virtual void Value
          (const Vector_<>& x,
25         Vector_<>* value)
       const = 0;

       virtual Vector_<x_dep_t> XDependence() const = 0;
   };
30
   class ScalarCoeff_ : public Coeff_
   {
   public:
```

```
    virtual void Value
35    (const Vector_<>& x,
        double* value)
    const = 0;

    virtual x_dep_t XDependence() const = 0;
40 };
```

The `Value` functions take an output parameter to be written on, rather
than returning a result, which violates our normal preference for functional
code; the repeated construction of vectors and matrices would be too slow.
For the scalar, while there is no efficiency penalty, it seems worthwhile to
keep the function signatures consistent.

These calculators are produced by the model, and know nothing of
coordinate transformations. Thus they work completely in x-space. We
also make available to all models the utilities:

```
─────────────────────────── PDE.h ───────────────────────────
namespace PDE
{
    MatrixCoeff_* NewConstCoeff(const Matrix_<>& val);
    VectorCoeff_* NewConstCoeff(const Vector_<>& val);
5   ScalarCoeff_* NewConstCoeff(double val);
}
```

7.9.4   *Forward Induction*

The persistence of fully-explicit, usually trinomial, "trees" for solution of
PDEs is a mixture of tragedy and farce. Even the most naive extension of
Crank-Nicholson – with cross terms computed using a fully explicit method
– is stable and substantially more accurate than even a carefully built tree.
Trees may be useful in pedagogy, but have no place in the working world
of pricing.

A common excuse given by practitioners of this sorry art is that implicit
or semi-implicit methods, despite their other desirable properties, do not
support forward induction. Forward induction is doubtless useful: it is
needed to fit the discount curve in some models such as Black-Karasinski,
and is used for pricing many options in a single PDE sweep in Dupire-style
local vol models. But forward induction is far less difficult than is often
thought, even for sophisticated PDE solvers.

Suppose we have a set of possible states x_i, and corresponding values $U_i^{(+)}$ at time t_+.[16] The task of a backward solver is to compute $U_i^{(-)}$ at some earlier time t_-; all the solvers we consider are linear in U, so we have $U_- = MU_+$ for some matrix M independent of U.

Next, let $G_i^{(\pm)}$ represent the state prices at time $t_\pm$. We know the present value of the payout is $G^{(-)} \cdot U^{(-)} = G^{(+)} \cdot U^{(+)}$; and this is true for any payout U. Thus $G^{(-)} \cdot MU^{(+)} = G^{(+)} \cdot U^{(+)}$ for any $U^{(+)}$, which requires $G^{(+)} = M^T G^{(-)}$.

For models with space-dependent coefficients, the forward Kolmogorov equation is complicated by many extra terms, and if we discretize it we cannot take advantage of the above relation. Separately discretizing the forward equation is a mistake; in fact, even *deriving* the forward Kolmogorov equation is a mistake. We should discretize the backward equation *only*, and then take the "numerical dual" using the above relation. (This does require some judgement in the choice of a boundary condition which is sensible for both directions; however, bear in mind that the forward evolution only makes explicit the boundary's preexisting effects on backward-induction pricing.) The result is that the forward step is as simple to implement as the backward – in fact, they share the vast majority of their code – and the resulting state prices are consistent with backward induction pricing to machine precision.

7.10 American Monte Carlo

In order to simulate optimal exercise of a Bermudan option, we must make some statement about the future expected value along two available branches; usually one branch's value is a known exercise value, and we are concerned with the *continuation value* along the other branch.

In non-recombining or "bushy tree" methods, we essentially spawn a new Monte Carlo beginning at the node where we must make an exercise decision. This child simulation must spawn its own children when it reaches a later exercise date; thus the cost of simulation grows exponentially with the number of exercise dates, relegating this method to the world of thought experiments.

American Monte Carlo (AMC) methods seek to replace the computation of continuation values using child simulations with an estimate gleaned from the paths we have already run. Thus AMC is fundamentally about

---

[16]Our notation is that of a single spatial dimension, but our analysis is not restricted to that case: the index i here simply enumerates all nodes in any number of dimensions.

estimation of continuation values, and only incidentally about *optimization* of exercise decisions. As such, it relies on the independent variables of the estimator, which must be easily measurable at each node of the simulation; these are called *observables*. In practice we prefer to estimate the *exercise gain*, which is simply the difference between the exercise and continuation values.

7.10.1 *Recursive Partitioning*

The well-known Longstaff-Schwartz AMC algorithm uses a very simple estimator, relying on a rich set of observables to achieve acceptable results. We can reduce the observables, thus simplifying the setup and computation at each node and reducing the coupling between AMC and model choice, by using a more sophisticated estimation procedure.

One way to improve the estimator is to increase its locality through recursive partitioning of the path set, or "bundling." Here we sort according to the last observable, partition the path set into bundles, then use the next-to-last observable to sort and partition into smaller bundles, and so on. The algorithm proceeds back-to-front in order to make the resulting bundles more local in the observables nearest the front, effectively making those observables more important.

The partitioning code itself creates two outputs, a `key` by which the paths are ordered and a list of `breakpoints` within the key; thus `key[0]`, `key[1]`, ... `key[b-1]` are the indices of the paths in the first bundle, where b is the first breakpoint.

```
                          ───────── AMC.cpp ─────────
    void Partition
        (const Matrix_<>& observables,
         const Vector_<int>& n_bundles,
         bool bundle_first,
5        Vector_<int>* key,
         list<int>* breaks)
    {
         const int nPaths = observables.Columns();
         *key = IndicesTo(nPaths);    // [0, nPaths)
10
         breaks->clear();
         breaks->push_back(0);
         breaks->push_back(nPaths);
         const int minD = bundle_first ? 0 : 1;
15       for (int d = observables.Rows() - 1; d >= minD; --d)
         {
             ObservableLess_ compare(observables, d);
```

```
     int from = breaks->front();
     for (auto p = Next(breaks->begin());
20         p != breaks->end(); from = *p++)
     {
         sort(&(*key)[from], &(*key)[*p], compare);
         const int size = *p - from;
         const int nb = Min(size, n_bundles[d]);
25       for (int j = 1; j < nb; ++j)
             breaks->insert(p, from + (size * j) / nb);
     }
   }
 }
```

This relies on the sorting predicate:

```
———————————————————— AMC.cpp ————————————————————
struct ObservableLess_
{
    const Matrix_<double>& obs_;
    const int depth_;
5   ObservableLess_(obs_, depth_);
    bool operator()(int lhs, int rhs) const
    {
        return obs_(depth_, lhs) < obs_(depth_, rhs);
    }
10 };
```

7.10.2 *Biases*

The flag `bundle_first` above is useful because, once we have sorted and
partitioned using all the other observables, we can use other methods for
the one-dimensional problem within a single bundle. In particular, a nonparametric fit based on linear smoothing splines can capture more of the
signal than a rigid estimator across bundles, while simultaneously enlarg-
ing the set of paths influencing the estimator; that is, it can reduce both
granularity bias from the nonzero spatial extent of the basis of estimation
and *small-sample bias* from the finite pathset size.

There is a third, more vicious bias with which we must also contend;
the *lookback bias* resulting from the influence of a path's own future on its
exercise decision. If the bundles are too small or the smoothing spline too
flexible, the path's observed future in this particular pathset will dominate
our estimate of the expected continuation value, and the prices we generate
will be those of lookback options.[17] This leads us to consider the *jackknifed*

---

[17]Worse, in fact, since the payout need not be discounted over a lookback period.

estimator in which a path's own future is excluded from the estimation process for that path; however, this is not a panacea, since it does nothing to control the other two (downward) biases.

Consider a simplified model of the estimation procedure. For a given path, let μ be the true expected gain from exercise, and assume that our measurement of the gain (*i.e.*, the observed gain from the future of that path) is normally distributed. Thus our measured gain on a given path is $\mu + \epsilon_i$ where the ϵ's are i.i.d. $N(0,1)$ deviates. Without loss of generality we can assume the measurement variance is 1. We also assume the pathset is large so that we can ignore edge effects; this lets us simplify our notation by considering "path zero" within a pathset extending to both positive and negative indices.

The value realized on the path, including the option to exercise, can be written in terms of this error term as $(\mu + \epsilon_0)\mathbf{1}_{\hat{\mu}_0 > 0}$ where $\hat{\mu}_0$ is the estimated exercise gain on this particular path. If our estimator is linear, then in our model we can write it as $\mu + \beta\epsilon_0 + \gamma\epsilon_\perp$, where $\epsilon_\perp$ is uncorrelated with ϵ_0 and encapsulates the total effect of all other paths. This expectation can be computed with a simple rotation of variables, obtaining

$$\mu N\left(\frac{\mu}{\sigma}\right) + \frac{\beta}{\sigma}\phi\left(\frac{\mu}{\sigma}\right)$$

where ϕ is the normal density function and $\sigma^2 \equiv \beta^2 + \gamma^2$. The true value is of course μ^+; the former term is always less than this due to small-sample bias, while the latter positive term is the lookback bias.

Since we are interested in the total bias over the whole range of paths, we next integrate over μ. The result is $\beta - \frac{1}{2}\sigma^2$ – to which *the total bias introduced into pricing* is proportional. This shows the possibility of balancing the lookback bias against the small-sample bias, especially in the region where k is significantly below 1, where both increase in importance.

If we further assume that there is only one observable, and that the paths are equidistant in this independent variable, then our linear smoothing spline reduces to an average weighted by a double-exponential with decay constant k:

$$\hat{y}_0 = \mu + \frac{1-k}{1+k}\sum_{i \in \mathbb{Z}} \epsilon_i k^{-|i|} = \mu + \frac{1-k}{1+k}\epsilon_0 + \frac{k}{1+k}\sqrt{\frac{2(1-k)}{1+k}}\epsilon_\perp.$$

For brevity we define $H \equiv (1-k)/(1+k)$, so in our notation

$$\beta = H, \qquad \gamma = (1-H)\sqrt{\frac{H}{2}};$$

similarly, the jackknifed estimator has

$$\beta = 0, \qquad \gamma = \sqrt{\frac{H}{2}}.$$

Now suppose that we use a linear combination of the two estimators, weighting the first by w_L (the *lookback weight*). The combined estimator has

$$\beta = w_L H, \qquad \gamma = (1 - w_L H)\sqrt{\frac{H}{2}};$$

the bias-cancelling condition becomes

$$w_L = \frac{1 - \sqrt{\frac{2}{2+H}}}{H},$$

which rapidly approaches $w_L = \frac{1}{4}$ as $H \to 0$ (the limit of rigid smoothing splines).

Thus, by the fairly simple trick of averaging the raw and jackknifed estimators, we produce a combined estimator which causes the worst upward and downward biases in pricing to cancel. While this result has been derived under highly restrictive assumptions, in practice it is applicable to real problems and greatly increases the power of estimation-based deciders. The resulting American Monte Carlo pricing is powerful enough for any product for which we can choose a sensible set of observables, and has been used to run a large book of callable exotics using multi-factor models.

This page intentionally left blank

Chapter 8

Schedules

When we attempt to apply mathematical techniques to financial instruments, there is a constant tension between the "pure" mathematical regime – of continuously compounded rates, uniform daycounts, and so forth – and the idiosyncratic real world, with its profusion of entrenched conventions. This is particularly pronounced in fixed income, but is an issue for any underlying. The unifying goal of schedule code is to bridge this divide by providing to the mathematical side the tools to encapsulate these conventions and quarantine them so we can get on with our real work. We make no pretense that schedules are interesting, but they are crucial.

8.1 Enumerated Switches

The details of a schedule-based contract, such as a Libor swap, are determined by a combination of choices: roll convention, daycount, and so forth. For example, the fixed coupon rate is converted to a cash payment amount for one period by multiplying by the notional amount and by a daycount fraction, which is (probably) computed by applying some accepted "daycount basis" to the accrual period start and end dates. Thus our parametrization of a swap must indicate the method for this computation.

This calls for a machine-generated `enumeration` type as in Sec. 3.8 – a class constructible from a string. But there is a complication: we must be prepared for new methods to be added. For instance, when we expand our emerging-markets operations we must begin dealing with Brazilian swaps whose daycount is 1/252 of the number of Rio de Janeiro business days in the interval. Machinist supports code generation for such an *extensible enumeration*, triggered by adding the keyword `extensible` in the mark-up:

```
———————————————— DayBasis.h ————————————————
/*IF--------------------------------------------------------
enumeration DayBasis
    Daycounts for accrued interest
extensible
alternative ACT_365F ACT/365F ACT_365FIXED ACT/365FIXED
    Always uses a 365-day year
alternative ACT_365L ACT/365L ISMA_Year
    Sometimes uses a 366-day year
alternative ACT_360 ACT/360 MONEY ACTUAL/360
alternative ACT_ACT ACT/ACT ACTUAL/ACTUAL
alternative BOND 30_360 30/360 30_360_US
method double operator()(const Date_& start_date, \
const Date_& end_date, const DayBasis::Context_* context) const;
-IF--------------------------------------------------------*/
```

Because we have not declared this type to be `switchable`,[1] the generated code will look somewhat different from that of the enumerations shown in Sec. 3.8.

```
———————————————— MG_DayBasis_enum.h ————————————————
class DayBasis_
{
    enum class Value_ : char
    {
        _NOT_SET=-1,
        ACT_365F,
        ACT_365L,
        ACT_360,
        ACT_ACT,
        BOND,
        _EXTENSION,
        _N_VALUES
    } val_;

public:
    class Extension_
    {
    public:
        virtual ~Extension_();
        // Must implement DayBasis_ interface
        virtual const char* String() const = 0;
        virtual double operator()(const Date_& start_date,
                const Date_& end_date,
                const DayBasis::Context_* context) const = 0;
    };
```

---

[1]We could define a `closed` tag, to label enumerations which are neither switchable nor extensible, but this is seldom useful.

```
   private:
       Handle_<Extension_> other_;
       const Extension_& Extension() const;
       DayBasis_(val_) {assert(val < Value_::_EXTENSION);}
30     DayBasis_(const Handle_<Extension_>& imp);
       friend bool operator==(const DayBasis_&, const DayBasis_&);
       friend struct ReadStringDayBasis_;
       friend Vector_<DayBasis_> DayBasisListAll();
       friend bool operator<(const DayBasis_&, const DayBasis_&);
35 public:
       explicit DayBasis_(const String_& src);
       const char* String() const;
       DayBasis_() : val_(Value_::_NOT_SET) {}
       // idiosyncratic (hand-written) members:
40     double operator()(const Date_& start_date, const Date_&
               end_date, const DayBasis::Context_* context) const;
   };
```

Here the interface generator has to be a bit sneaky to generate the
`Extension_` base class: it notices that one of the members ends with `const`;
and uses that information to generate a virtual member of `Extension_`. We
need this extra step to avoid machine-generating invalid code like:

```
virtual typedef double result_type = 0;
```

If we are determined to avoid this hack, we can require the mark-up to state
explicitly which members are not to be transplanted to the `Extension_`
(*e.g.*, by marking them as `final` rather than simply `method`); Machinist
does not presently support this.

8.1.1 *Groundwork for Extensibility*

We must implement a constructor from string for `DayBasis_`, but we cannot
yet know what other values may be declared elsewhere. Also, in Sec. 3.8,
we have rashly promised to `ListAll` the values.

Thus we must supply a run-time registry for recognition functions and
the corresponding canonical names. As usual, we use Meyers singletons:

```
────────── MG_DayBasis_enum.inc ──────────
namespace
{
    Vector_<DayBasis_>& TheDayBasisList()
    {
5       RETURN_STATIC(Vector_<DayBasis_>);
    }
    map<String_, Handle_<DayBasis_::Extension_>>&
    TheDayBasisExtensions()
    {
```

```
10      RETURN_STATIC(map<String_, Handle_<DayBasis_::Extension_>>);
     }

}    // leave local
```

TheDayBasisExtensions is the registry to which extension classes will add themselves. Note that this implementation assumes a limited set of exact (modulo case, whitespace and underscores) matching strings for each enumerated value, rather than a more generalized notion of patterns. This is by design: the risk of false positives makes uninhibited pattern-matching too dangerous for enumerated switches in production.

```
                     ── MG_DayBasis_enum.inc ──
void DayBasis::RegisterExtension
   (const Vector_<String_>& names,
    const Handle_<DayBasis_::Extension_>& imp)
{
5    REQUIRE(TheDayBasisList().empty(), "Can't register a new "
            "DayBasis after enumerating all values");
     assert(!imp.Empty());
     DayBasis_RejectDuplicate(imp->String());
     for (auto pn = names.begin(); pn != names.end(); ++pn)
10   {
         DayBasis_RejectDuplicate(*pn);
         TheDayBasisExtensions()[String::Condensed(*pn)] = imp;
     }
     // check that two-way string conversion works
15   DayBasis_ check(imp->String());
     REQUIRE(check.String() == imp->String(),
            "String representation reconstructs the wrong "
            "DayBasis -- presumably a name clash");
}
```

The final check ensures that the **String** method implemented by **val** returns a string from which the same **DayBasis_** can be reconstructed; the initial call to **RejectDuplicate** assures us that no other **val** can intercept the strings we are looking for.

Registration depends on a supporting function, which ensures that the meaning of any previously valid string will not be changed:[2]

```
                  ── MG_DayBasis_enum_public.inc ──
void DayBasis_RejectDuplicate(const String_& test)
{
     static const ReadStringDayBasis_READ_FIXED;
     DayBasis_::Value_val;
5    NOTICE(test);
     REQUIRE(!READ_FIXED(test, &val),
```

---

[2]This replaces the awkward **MustFail** function in the first edition, which had the undesirable property of throwing exceptions during normal initialization.

```
        "Attempt to change meaning of fixed DayBasis string");
     REQUIRE(!TheDayBasisExtensions().count(String::Condensed(test)),
        "Attempt to change meaning of DayBasis string");
10  }
```

External users may wish to list all possible day bases at runtime – *e.g.*, to populate a drop-down list in a user interface. Our support for this is based on the function:

```
─────────────────── DayBasis.cpp ───────────────────
Vector_<DayBasis_> DayBasisListAll()
{
    static const Vector_<DayBasis_::Value_> ALL_BASE =
        { DayBasis_::Value_::ACT_365F,
5         DayBasis_::Value_::ACT_365L,
          DayBasis_::Value_::ACT_360,
          DayBasis_::Value_::ACT_ACT,
          DayBasis_::Value_::BOND };

10   if (TheDayBasisList().empty())
     {
         Vector_<DayBasis_> retval;
         set<String_> exists;
         for (auto b : ALL_BASE)
15       {
             retval.push_back(DayBasis_(b));
             exists.insert(retval.back().String());
         }
         for (auto e : TheDayBasisExtensions())
20       {
             if (!exists.count(e.second->String()))
             {
                 retval.push_back(DayBasis_(e.second));
                 exists.insert(e.second->String());
25           }
         }
         TheDayBasisList().Swap(&retval);
     }
     return TheDayBasisList();
30  }
```

The registration function will fail if this list has already been populated, since we cannot know what clients might otherwise be left holding an incomplete list.

We expect that registration will happen during an initialization phase, not as a result of interaction with the user.

All the above code is in fact machine-generated. The template used in the generation process is essentially a block of pseudocode, into which we substitute the string "DayBasis".

8.1.2 *30E/360 ISDA, ACT/ACT ISMA*

Some conventions do not fit naturally into this framework. 30E/360 ISDA, used by some German bonds, treats the last period differently from others: a period ending on the last day in February will be treated as February 30 for daycount purposes, *unless* it is the last coupon period. In actual/actual ISMA, used by some U. S. Treasury bonds, the period daycount cannot be determined without reference to the nominal start and end dates and the coupon frequency.

If we are to support these conventions, we have two unpalatable choices: we can replace the simple `operator()` above with

```
———————————————— DayBasis.h ————————————————
method double operator()(const Date_& start_date, const Date_& \
end_date, bool is_last, const Date_& nominal_start, const Date_& \
nominal_mat, int cpn_months) const;
```

and then ensure that all daycounts have this information whenever they might need it. Alternatively, we might treat legs using these daycount conventions as nonstandard, and supply separate special-purpose constructors for them. We reluctantly adopt the first approach.

We sweeten the bitter pill somewhat by packaging the otiose information in a single argument,

```
———————————————— DayBasis.h ————————————————
struct DayBasisContext_
{
    bool isLast_;
    Date_ nominalStart_;
5   Date_ nominalEnd_;
    int couponMonths_;
};
```

Now we can write the less voluble

```
———————————————— DayBasis.h ————————————————
method double operator()(const Date_& start_date, const Date_&\
end_date, const DayBasisContext_* context) const;
```

8.1.3 *BUS/252*

A similar problem is introduced by the presence of daycounts depending on a number of business days, which in turn depends on the holiday calendar used. This can be implemented in two ways: by making the daycount depend on an input holiday calendar (adding yet another argument to `operator()` above), or by observing that different holiday calendars create different day bases. We prefer the latter approach, which separates the day basis from the other uses of the holiday calendar in constructing a swap schedule. This obliges us to create a new day basis for each holiday schedule used with a Business/252 basis; however, such daycounting is widespread only in Brazil.[3]

We describe this additional day basis in another mark-up block:

```
―――――――――――――― Bus252.DayBasis.enum.if ――――――
alternative BRL_BUS_252 Brazil_BUS_252
'Brazilian daycount convention.
method double operator()(const Date_& start_date, const Date_&\
end_date, const DayBasisContext_* context) const;
```

The `method` must be repeated because the code generator will not have access to the markup for `DayBasis` itself. Now we have supplied enough information to generate code defining the extension class. We can place it completely in the C++ source file, and no protection is necessary, since it will be accessed only through the base class interface.

```
―――――――――――――― Bus252.cpp ――――――
struct Bus252_DayBasis_ : DayBasis_::Extension_
{
    enum class Value_ : char
    {
5       BRL_BUS_252,
        _Bus252_N_VALUES
    } val_;
    Bus252_DayBasis_(val_) {}

10  const char* String() const override;
    double operator()(const Date_& start_date,
        const Date_& end_date,
        const DayBasis::Context_* context) const override;
};
```

The registration is accomplished by our usual method, declaring a file-scoped static object whose constructor calls the registration function. Because Meyers singletons are initialized on the first call to the function containing their definition, we are not at risk of initialization-order bugs: the central registry will be created when it is needed.

```
―――――――――――――― Bus252.cpp ――――――
RUN_AT_LOAD(
    Handle_<DayBasis_::Extension_> h
        (new Bus252_DayBasis_
```

―――――――――――――――――――――――――
[3]Where it is a legacy of hyperinflation.

```
                    (Bus252_DayBasis_::Value_::BRL_BUS_252));
5    DayBasis::RegisterExtension
            ({ "BRL_BUS_252", "Brazil_BUS_252" }, h);
);
```

The body of `String()` is easy to machine-generate as well. The only task that remains for us is to implement the relevant `operator()`; this functionality will then be available to any user, once the code is loaded.

The calling code to use an enumerated type does not change when that type is made extensible; this is a crucial advantage, letting us open up our system in ways we may not originally have anticipated. We should as a rule avoid making enumerations `switchable`, but that is simply the good coding practice of minimizing unnecessary assumptions.

8.1.4 *Other Enumerations*

Machinist's `enumeration` will be our preferred method for defining any list of user choices. The uniform interfaces of the machine-generated code support the reliable generation of code in other areas – e.g., at the public interface, where we will take inputs of type `enum`. We will also enjoy the fringe benefits of code generation, particularly simultaneous generation of user help pages so there are no hidden "magic words."[4]

Though we have focused here on `DayBasis_`, other enumerated switches follow precisely the same pattern. Our job is to distill the use of the enumeration into a few C++ function signatures, which are declared as `methods` in mark-up descriptions and implemented by hand. Common enumerated types for schedule generation include

- Business-day convention (applied if a nominal period end is not a business day). This presents one subtlety, like that in the discussion of 30E/360 day basis above; sometimes accrual dates are not adjusted, though payment dates always are. Is this a feature of the leg, or of the roll convention?
- Roll type (for swaps whose period end dates are not computed from the start date, such as IMM swaps).
- Frequency, or period length. Because these are two forms for the same information, conversion to and from integers must be very explicit (does 1 mean monthly or annual)?
- Averaging frequency, for averaging basis swaps.

---

[4] A machine-generated help page for an `extensible` enum will inevitably be incomplete; we need to link to a loaded instance of the library to know what extensions are available.

8.2 Holidays

8.2.1 *Cities*

Each jurisdiction in which a trade might be consummated will define a *holiday calendar*, indicating which days are not *good business days* there. A jurisdiction, in this role, is called a *holiday center* or simply a *city*.

The set of holidays for a city is subject to change, sometimes at fairly short notice, as new special occasions are deemed worth celebrating. Also, an expanding international business will routinely find itself involved in new jurisdictions, necessitating the addition of new holiday calendars. Maintenance of these holiday calendars is not a natural or rewarding job for front-office quants.

Thus holidays are best regarded as configuration data, which we will read at library load time. The data are fairly simple: for each of some (fairly large) set of city codes, we will supply a set of all holidays over some (fairly wide) date range, probably from `Date::Minimum()` through `Date::Maximum()`. Holidays may be more compactly represented by rules (*e.g.*, 25 December is always a London holiday when on a weekday), but this rule-based representation is best kept encapsulated within the configuration loader, or within a completely separate application which generates data to be loaded. For our purposes, we will expect the library to obtain a `map<String_, Vector_<Date_>>`.

For efficiency, we will store this data in two parts: a map of cities to integer indices, and an index vector of holidays.

```
──────────────── HolidayData.cpp ────────────────
struct HolidayCenterData_
{
    String_ center_;
    Vector_<Date_> holidays_;
    HolidayCenterData_(center_, holidays_) {}
};
struct HolidayData_
{
    Vector_<Handle_<HolidayCenterData_>> holidays_;
    map<String_, int> centerIndex_;

    bool IsValid() const;
    void Swap(HolidayData_* other);
};
```

This representation (in a local namespace) lets us convert back and forth between cities and their indices. The `IsValid` code checks class invariants, such as correspondence between the keys of the `centerIndex_` and the `centers_`.

```
─────────────── HolidayData.cpp ───────────────
void Holidays::AddCenter(const String_& city,
    const Vector_<Date_>& holidays)
{
    assert(TheHolidayData().IsValid());
5   assert(ContainsNoWeekends(holidays));
    assert(IsMonotonic(holidays));
    NOTICE(city);

    HolidayData_ temp(CopyHolidayData());
10  REQUIRE(!temp.centerIndex_.count(city), "Duplicate holidays");
    temp.centerIndex_[city] = temp.holidays_.size();
    temp.holidays_.push_back
        (make_shared<const HolidayCenterData_>(city, holidays));

15  LOCK_DATA;  // mutex
    TheHolidayData().Swap(&temp);
    assert(TheHolidayData().IsValid());
}
```

Use of the copy-swap idiom protects us from the unlikely event of an exception during the insertion process, which would make the library unusable.

8.2.2 *Holiday Sets*

It is quite common for dates to depend on multiple holiday schedules. For instance, the Libor start date for a given fixing date uses both New York and London holidays; and an offshore corporate counterparty to a swap is likely to ask that swap payments be rolled to good business days in their city of operations.

Thus the fundamental object used in schedule generation is a *holiday set*, which can in special circumstances be empty, of holiday centers. This is constructed from a list of holiday center codes; we prefer a space-separated list (which can then be included in a larger comma-separated list on occasion), but our parser will accept commas as well, and possibly other separators. A blank string should not be interpreted as "no holidays", which is really a rare usage and invalid in many situations; instead, some special string like `"NONE"` should be required. For the same reason, we do not provide the holiday set with a default constructor.

A full holiday calendar stores the data from a list of centers:

```
———————————————— Holiday.h ————————————————
class Holidays_
{
    Vector_<Handle_<HolidayCenterData_>> parts_;
    friend class CountBusDays_;
public:
    Holidays_(const String_& src);
    String_ String() const;
    bool IsHoliday(const Date_& date) const;
};
```

The implementation of `IsHoliday` proceeds as one would expect:

```
——————————————— Holiday.cpp ———————————————
bool Holidays_::IsHoliday(const Date_& date) const
{
    for (auto ps : parts_)
        if (BinarySearch(ps->holidays_, date))
            return true;
    return false;
}
```

We move the functionality needed for `String` to a free function, so that it can also be called from the constructor. This function will give a reproducible output for any set of holiday centers only if they are unique and predictably ordered. We accomplish this by always calling `Unique` (see Sec. 4.2) on the center names before getting the corresponding `HolidayCenterData_`.

```
——————————————— Holiday.cpp ———————————————
String_ NameFromCenters
    (const Vector_<Handle_<HolidayCenterData_>>& parts)
{
    static const auto ToName = []
        (const Handle_<HolidayCenterData_> h) {return h->center_;};
    return String::Accumulate(Apply(ToName, parts), " ");
}
```

By implementing `Holidays_` in the same source file as `AddHolidayCenter`, we ensure that `TheHolidayData` is accessible from both while remaining insulated from non-holiday code.

More complex tasks, like a count of business days in an interval, can in principle be implemented as nonmember functions calling `IsHoliday` repeatedly. For better performance, we may wish to support this more directly. The ability to count holidays in an interval could be supported by pre-merging the holiday lists in a new member `vector<Date_>`

`Holidays_::vals_:`

```
// deprecated version
int Holidays_::CountHolidays
    (const Date_& begin, const Date_& end) const
{
    auto pf = LowerBound(vals_, begin);
    return lower_bound(pf, vals_.end(), end) - pf;
}
```

This counts the elements of `vals_` lying in the half-open interval [begin, end), echoing the standard use of those names. Note the use of both the standard and the container-level versions of lower bound, as explained in Sec. 4.2.

But populating `vals_` on construction is likely a pessimization; it requires at least a vector copy for each holiday center in the set. We could make `vals_` mutable, and ensure it is initialized at the start of `CountHolidays`: this is less inefficient but reflects the poor practice of mixing states (*i.e.*, initialized and uninitialized) within the same type; see Sec. 2.3.

Since there is no requirement that a `HolidayCenterData_` correspond to a single city, we can store additional data for a lazily computed set of city combinations. We create a separate class for the task of counting business days:

Holiday.h

```
class CountBusDays_
{
    Holidays_ hols_;
public:
    CountBusDays_(hols_);
    int operator()(const Date_& begin, const Date_& end) const;
};
```

Now, instead of

```
Holidays::CountBusDays(holidays, begin, end)
```

we will write

```
CountBusDays_(holidays)(begin, end)
```

which is not substantially more complex.

So far, it does not look as if anything has been gained. However, the constructor of `CountBusDays_` can create and store an optimized holiday calendar:

```
                      ── Holiday.cpp ──
CountBusDays_::CountBusDays_(const Holidays_& src) : hols_(src)
{
    if (src.parts_.size() > 1)  // the holiday calendar is not
    merged
    {
        LOCK_COMBOS; // mutex
        auto& combo = TheCombinations()[src.String()];
        if (combo.Empty())
        {
            Vector_<Date_> merged;
            for (auto p : src.parts_)
                merged.Append(p->holidays_);
            combo.reset(new HolidayCenterData_
                (src.String(), Unique(merged)));
        }
        hols_.parts_ = Vector::V1(combo);
    }
}
```

Here we have assumed access to the internals of `Holidays_`; thus
`CountBusDays_` should be a **friend** of that class.[5]

This is the main time-intensive application of holiday schedules. We
could in principle optimize other functions (like finding the next good business day) in the same way, but there is no real gain from this. It is better
to keep the classes simple and supply nonmember functions as needed.

We can improve the constructor of `Holidays_` to use the precomputed
information when it is available:

```
                      ── Holiday.cpp ──
Holidays_::Holidays_(const String_& src)
{
    Vector_<String_> hc = String::Split(src, ' ', false);
    parts_ = Unique(Apply([](const String_& c){return
        Holidays::OfCenter(Holidays::CenterIndex(c)); }, hc));
    if (parts_.size() > 1)
    {
        LOCK_COMBOS;  // mutex
        auto cd = TheCombinations().find(NameFromCenters(parts_));
        if (cd != TheCombinations().end())
            parts_ = Vector::V1(cd->second);
    }
}
```

This constructor first splits the input string (on space) into a set of cities

---

[5]Friend status is not completely necessary; we could index `TheHolidays` by a string key,
obtained from `src.String`, with only a minor efficiency penalty.

and gets the holiday center data for each; then, if more than one center is present, it checks for an already-optimized combination (but does not create one). Once `CountBusDays_` has been called for a given city combination, the merged holiday calendar will be computed and cached, so all holiday calendars constructed thereafter will enjoy a highly efficient implementation.

8.3　Currencies

Besides being the means of payment, currencies also supply default conventions for many trades: once I know that a Libor swap is denominated in USD, I can surmise that its fixed side is semiannual with 30/360 day basis and New York payment holidays, and so on. Of course, these may be overridden for a particular trade, but it is necessary to know the defaults: they complete the definition of market observable indices and of quoted yield curve build instruments, since the markets for nonstandard trades are much less deep, less liquid and less public.

For every currency, we must be able to look up a wide variety of facts. Since these facts are of different types, from `int` on up, storing and indexing them is a surprisingly deep problem. We must decide on

- Order of indexing – is the currency the accessor's first or last argument?
- Extensibility – can we store and fetch conventions whose existence was not foreseen?
- Typing – where does the conversion from text data occur?
- Syntax – can we fetch a single fact without extra type clutter?

The syntax and typing constraints turn out to be surprisingly strong. To have a clean calling syntax, we must type some permutation of `Ccy::Conventions`, `"USD"`, `SwapPayHolidays` and get back a `Holidays_` without any further ado. Thus the type information must come from the name of the thing requested: it must identify a member datum, function, or concrete class.

The desire for extensibility dictates that the currency should be the last, not the first, argument; then new quantities can be introduced by introducing new functions in a namespace, which can be done anywhere. This also improves physical code structure, since different conventions need not all be tied together.

Thus we end up writing:

```
Ccy::Conventions::SwapPayHolidays()("USD");
```

which is not so different from our first guess.

The return value of `SwapPayHolidays` should expose a writing interface as well (otherwise we must create a separate writing interface elsewhere). This leads us to the following implementation:

```
_____ Facts.h _____
template<class K_, class V_> class OneFact_ : public FactBase_
{
public:
    virtual const V_& operator()(const K_& key) const = 0;

    class Writer_ : noncopyable
    {
    public:
        virtual void SetDefault(const V_& val) = 0;
        virtual void operator()(const K_& key, const V_& val) = 0;
    };
    virtual Writer_& XWrite() const = 0;
};
```

`FactBase_` simply provides a virtual destructor to the template class.

Thus we store facts with code like:

```
Ccy::Conventions::LiborFixDays().XWrite().SetDefault(2);
Ccy::Conventions::LiborFixDays().XWrite()(Ccy_("GBP"), 0);
```

Such setup code should be collected in an initialization module, giving us some future control over the storage of facts – configuration, database, web service, etc.

8.3.1 *Internals*

The above interface will clearly be supported by a singleton data holder, a dictionary plus default value, whose definition can be localized to a single source file.

```
_____ CurrencyData.h _____
template<class T_> struct CcyDependent_
{
    Handle_<T_> background_;
    map<String_, Handle_<T_>> specific_;
};
```

We hold the `specific_` values in `Handle_`s so that our code will work for classes which lack a default constructor.

Our implementation of a currency-dependent fact will store a `CcyDependent_`. We implement the writer by providing access directly to this data:

```
——————————— CurrencyData.cpp ———————————
template<class T_> class XFactWriterImp_
       : public Ccy::Fact_<T_>::Writer_
{
    CcyDependent_<T_>& data_;    // we do not own
5  public:
       XFactWriterImp_(CcyDependent_<T_>& d) : data_(d) {}

       void SetDefault(const T_& val)
       {    data_.background_.reset(new T_(val));    }
10     void operator()(const Ccy_& ccy, const T_& val)
       {    data_.specific_[ccy].reset(new T_(val));    }
};
```

Here `Ccy::Fact_` is a template typedef for a fact with a `Ccy_` key:

```
——————————— CurrencyData.cpp ———————————
namespace Ccy
{
    template<class T_> using Fact_ = OneFact_<Ccy_, T_>;
    // ...
```

We use a level of indirection to ensure that, when we construct our fact holder, the writer is constructed with a valid reference to the data.[6]

```
——————————— CurrencyData.cpp ———————————
template<class T_> class OneFactImp_ : public Ccy::Fact_<T_>
{
    CcyDependent_<T_> vals_;
    unique_ptr<typename Ccy::Fact_<T_>::Writer_> writer_;
5  public:
    OneFactImp_()
    {
        writer_.reset(new XFactWriterImp_<T_>(vals_));
    }
10
    const T_& operator()(const Ccy_& ccy) const
    {
      auto pc = vals_.specific_.find(ccy);
      if (pc != vals_.specific_.end())
15        return *pc->second;
      REQUIRE(!vals_.background_.Empty(),
```

---

[6]We could take the writer off the heap and rely on order of initialization, but the gains are insignificant.

```
              "No default for '" + String_(ccy.String()) + "'");
          return *vals_.background_;
      }
20    Writer_& XWrite() const { return *writer_; }
};
```

Now each function in `Ccy::Conventions` returns a Meyers singleton of this type. For each convention we need only one brief and simple function:

```
──────── CurrencyData.cpp ────────
#define SINGLETON_FACT_ACCESSOR(type, func)          \
const Ccy::Fact_<type>& Ccy::Conventions::func()     \
{ RETURN_STATIC(const OneFactImp_<type>); }

5  SINGLETON_FACT_ACCESSOR(Holidays_, SwapPayHolidays)
// ...
```

This code will be called at library load time by a configuration reader, which will read facts from some accepted source (*e.g.*, a database or configuration file) and write them into the store for that session. None of these methods are currency-specific; they will work equally well to store, *e.g.*, parameters of commodity futures contracts.

8.4 Increments

A good part of the fixed income business revolves around incrementing dates; by months, by calendar or business days, and so on. This is such a general concept that it is worth embodying in its own code.

```
──────── DateIncrement.h ────────
namespace Date
{
    class Increment_ : noncopyable
    {
5   public:
        virtual ~Increment_();
        virtual Date_ FwdFrom(const Date_& date) const = 0;
        virtual Date_ BackFrom(const Date_& date) const = 0;
    };
10
    Handle_<Increment_> ParseIncrement(const String_& src);
}

inline Date_ operator+(const Date_& d, const Date::Increment_& i)
15  {
    return i.FwdFrom(d);
```

```
}
inline Date_ operator-(const Date_& d, const Date::Increment_& i)
{
20    return i.BackFrom(d);
}
```

We distinguish two types of increment:

(1) Increment by some number of intervals;
(2) Increment to the next (or previous) of some set of dates.

The latter category includes, *e.g.*, advancing to the next quarterly IMM date. The best solution is an enumerated list describing which special dates allow this:

─────────────── *DateIncrement.cpp* ───────────────
```
/*IF---------------------------------------------------------
enumeration SpecialDay
    Date families supporting jump-to-next
alternative IMM IMM3 IMM_QUARTERLY
5       Quarterly IMM dates
alternative IMM1 IMM_MONTHLY
        Monthly IMM dates
alternative CDS CDS3 CDS_QUARTERLY
        Quarterly CDS standard maturities
10 alternative EOM
        End of a month
method Date_ Step(const Date_& base, bool forward) const;
-IF--------------------------------------------------------*/
```

Other sets of special days might be added, either in this file or in extensions (if we mark **SpecialDay** as **extensible**; see Sec. 8.1). The implementation of one type of increment is now straightforward:

─────────────── *DateIncrement.cpp* ───────────────
```
class IncrementNextSpecial_ : public Date::Increment_
{
    SpecialDay_ targets_;
    Date_ FwdFrom(const Date_& d) const override
5       { return targets_.Step(d, true); }
    Date_ BackFrom(const Date_& d) const override
        { return targets_.Step(d, false); }
public:
    IncrementNextSpecial_(targets_) {}
10 };
```

The code of **SpecialDays_::Next** will also be useful for Libor futures rates, allowing us to compute the IMM date in any given month by looking forward from the end of the previous month.

The other class of increment is also supported by an enumerated list, this time of step sizes:

```
/*IF----------------------------------------------------------  DateIncrement.cpp
enumeration DateStepSize
     Repeatable step length
alternative Y YEAR YEARS
alternative M MONTH MONTHS
alternative BD BUS_DAY BUSINESS_DAY
alternative CD CAL_DAY CALENDAR_DAY
method Date_ operator()(const Date_& base, bool forward, \
int n_steps, const Holidays_& holidays) const;
-IF---------------------------------------------------------*/
```

We use a boolean flag input to `operator()` to distinguish forward from backward steps, rather than a signed integer `n_steps`; this allows us to use `n_steps == 0` to mean forward or backward adjustment to a good business day (with no change if the `from` date is already a good business day). We might extend this enumeration with additional alternatives overriding the roll convention (*e.g.*, to allow rolling into a new month, which is nonstandard). Note that `DAY` or `D` are not valid steps; we must explicitly ask for calendar or business days.

An increment now specifies how many steps to take of a given type, and also gives the holiday calendar to use:

```
                        Increment.cpp
class IncrementMultistep_ : public Date::Increment_
{
     int nSteps_;
     DateStepSize_ stepBy_;
     Holidays_ hols_;

     Date_ FwdFrom(const Date_& d) const
     {
         return stepBy_(d, true, nSteps_, hols_);
     }
     Date_ BackFrom(const Date_& d) const
     {
         return stepBy_(d, false, nSteps_, hols_);
     }
public:
     IncrementMultistep_(nSteps_, stepBy_, hols_) {}
};
```

Now building a grammar for increment specification is very simple, because each increment either begins with a number (of steps to take), or is a `SpecialDays_`. We need to specify two features of the grammar; a separator between the `DateStepSize` and the (optional) `Holidays`, and a separator between multiple increments. The latter protects users from an unnecessary distinction between single and compound increments; for example, "the next London business day after the next New York and London business day after (a day)" is a reasonable specification of an increment, and we should treat it as such. We prefer ; for the former separator, and & for the second; thus we would write such an increment as `1BD;NY LON&1BD;LON`. Note the left-to-right application; date increments are not commutative.[7] A more sophisticated parser could allow omission of the &, accepting `"4Y9M"` as an equivalent to `"4Y&9M"`.

We support compound increments by defining a new type

```
──────────────── DateIncrement.cpp ────────────────
   struct IncrementCompound_ : Composite_<const Date::Increment_>
   {
       template<class F_> Date_ Apply(Date_ d, const F_& func) const
       {
5          for (auto inc : contents_)
               d = (inc.get()->*func)(d);
           return d;
       }
       Date_ FwdFrom(const Date_& date) const
10             { return Apply(date, &Increment_::FwdFrom); }
       Date_ BackFrom(const Date_& date) const
               { return Apply(date, &Increment_::BackFrom); }
   };
```

which lets user code effectively handle compound increments (such as `"4Y9M"` or `"5Y&CDS"`) without always thinking about their internals.

8.5 Legs

A *swap* is the repeated scheduled exchange of one kind of payment for another; but these exchanges often are not synchronous or even one-for-one. For example, the standard USD Libor swap exchanges quarterly Libor for a semiannual fixed payment. Thus the individual leg, not the swap, is the building block on which we should concentrate.

We must distinguish the schedule of payments from the determination of the payment amounts. The former is described by an *accrual schedule*, often loosely called a *leg schedule*, which is in turn a sequence of *accrual periods*:

---

[7]This is the usual cause of "Thursday bugs."

```
                    ──────────── AccrualPeriod.h ────────────
   struct AccrualPeriod_
   {
       Date_ startDate_;
       Date_ endDate_;
5      DayBasis_ couponBasis_;
       double dcf_;    // redundant but handy
       Handle_<DayBasis::Context_> context_;
       bool isStub_;    // used in forming Libor rates

10     ~AccrualPeriod_();    // "empty" but out-of-line
       AccrualPeriod_();
       AccrualPeriod_(startDate_, endDate_, couponBasis_);
       AccrualPeriod_
           (startDate_, endDate_, couponBasis_, context_, isStub_);
15 };
```

Now we must describe the *coupon rate*, which is multiplied by dcf_
to give the amount paid. This is a surprisingly nontrivial problem. We
might imagine a base class with a virtual CouponRate function; but this
would place the burden of rate computation at the very low level of leg
periods, not at the level of yield curves and models where it belongs. Also,
we cannot predict in advance all the methods for fixing a coupon rate,
nor the subsidiary data each will require; thus we need the flexibility of a
full-fledged object. This leads us to the somewhat unsatisying:

```
                    ──────────── CouponRate.h ────────────
   struct CouponRate_ : noncopyable
   {
       virtual ~CouponRate_() {}
   };
```

We will be obliged to construct free functions which discriminate different
types of CouponRate_ by dynamic_casting.

In particular, we will immediately define the common rate types:

```
                    ──────────── CouponRate.h ────────────
   struct FixedRate_ : CouponRate_
   {
       double rate_;
       FixedRate_(rate_) {}
5  };

   struct LiborRate_ : CouponRate_
   {
       DateTime_ fixDate_;
10     Ccy_ ccy_;
```

```
    String_ tenor_;
    LiborRate_(fixDate_, ccy_, tenor_);
};
```

We have chosen to make the `LiborRate_` contain its fixing date, rather than relying on obtaining it from the leg at the time of computation. This lets us make the rate computation fully independent. The `tenor_` should be parseable as a `DateIncrement_`.

Now we can build a leg by defining, for each period, the accrual terms and rate fixing terms:

```
─────────────────────── Period.h ───────────────────────
struct LegPeriod_
{
    Handle_<AccrualPeriod_> accrual_;
    Handle_<CouponRate_> rate_;
    Date_ payDate_;
};
```

This `struct` is designed to be held in vectors. The `accrual_` is held in a `Handle_` because `AccrualPeriod_` has no default constructor. We could provide a sorting relationship and use `set` instead of `Vector_`, but this has negative practical value. The periods of a single leg are generated in order and stay that way; and we will later see methods which combine multiple legs, without need for an ordering.

8.5.1 *Stubs*

For first- and last-period stub rates on Libor swaps, we must usually interpolate between published Libor fixings to generate the paid rate. If we write:

```
struct InterpLiborRate_ : CouponRate_ // deprecated
{
    Handle_<LiborRate_> r1_, r2_;
    double w1_;    // and w2 = 1-w1
};
```

then we have a special-purpose structure which still must be supported wherever `LiborRate_` is.

Instead, we should write:

```
─────────────── CouponRate.h ───────────────
struct SummedRate_  : CouponRate_
{
    Vector_<pair<double, Handle_<CouponRate_>>> rates_;
};
```

This supports rates with margin and float-float spreads as well, with no substantial complexity burden. We use a `Vector_` to hold any number of rates for interpolation, rather than restrict to exactly two; the likely alternative is the case of only one rate (*e.g.*, a rate transformed by a non-unit gearing). Also see Sec. 11.6 for more on queries of `CouponRate_`.

8.5.2 *Build from Parameters*

Leg schedules are not directly specified in the terms of a trade; they are worked out from higher-level parameters. These parameters, rather than the leg itself, should be `Storable_`.[8]

There is some question whether to store parameters which agree with the defaults for the leg's currency; for example, a USD Libor leg is quarterly by default, so should we store the coupon period for a quarterly floating leg? If we do so, then we are protected in case the default changes, since existing trades have all their details stored. But in storing these details, we will bloat the trade description; and, should we make any mistake as to conventions, this mistake will be enshrined in any booked trades. Here we adopt the policy of storing as little data as possible.

It is convenient to form a `Storable_` containing the information needed to form a swap leg. We describe its object layout in mark-up:

```
─────────────── LegParams.h ───────────────
/*IF--------------------------------------------------------
storable LegScheduleParams
        Parameters for the leg coupon and accrual dates
version 1
5  &members
name is ?string
        Name of the storable object
startDate is date
matDate is ?date
10        Fixed maturity date, if specified directly
tenor is ?string
        Leg tenor to determine maturity, if not specified directly
```

---

[8]By storing low-level information like payments without the higher-level information from which they are derived, we sever the link between the contents of our system and the real-world objects – trade confirmations – that they represent. Modification becomes perilous and migration impossible.

```
   couponPeriod is ?string
      A date increment string
15 dayBasis is ?string
      The daycount method for computing coupon amounts
   rollDay is ?integer
      A fixed day of the month for rolls, if specified
   rollDirection is ?string
20 rollSpecial is ?string
      E.g., for IMM rolls
   stubAtEnd is ?boolean
   longCoupon is ?boolean
   payUpfront is ?boolean
25 payHoidays is ?string
   payDelay is *integer
      A single offset (number of business days) for all payments,
      or a vector of one entry per payment
   &conditions
30 matDate_.IsValid() == tenor_.empty()\Specify mat date xor tenor
   -IF-------------------------------------------------------*/
```

If we should later need to change the contents (except by adding optional members), we must create a new mark-up block with a different **version** but also keep the old mark-up block for backward compatibility.

From the mark-up, Machinist generates a **Reader_** containing the above members:

```
                   ——— MG_LegScheduleParams_Read.inc ———
   namespace LegScheduleParams_v1
   {
      struct Reader_ : Archive::Reader_
      {
5        String_ name_;
         Date_ startDate_;
         Date_ matDate_;
         String_ tenor_;
         String_ couponPeriod_;
10       String_ dayBasis_;
         boost::optional<int> rollDay_;
         String_ rollDirection_;
         String_ rollSpecial_;
         boost::optional<bool> stubAtEnd_;
15       boost::optional<bool> longCoupon_;
         boost::optional<bool> payUpfront_;
         String_ payHoidays_;
         Vector_<int> payDelay_;
         // ...
```

Our **class LegScheduleParams_** will contain the same data, though our code generation scheme does not permit us to share the class definition.

Now we can create a generator for a fixed leg. We will collect such functions in **namespace LegBuild**.

```
───────── LegSchedule.h ─────────
Vector_<LegPeriod_> Fixed
    (const Ccy_& ccy,
     const LegScheduleParams_& schedule,
     const RecPay_& rec_pay,
5    const Vector_<>& notional,
     const Vector_<>& coupon,
     const NotionalExchange_* exchange = nullptr);
```

The **RecPay_** and **NotionalExchange_** classes are machine-generated enumerations, as in Sec. 8.1. The **notional** and **coupon** may be single numbers, or vectors with one entry per coupon period. The parameters of the schedule are taken from the input **schedule** when possible; if they are not present there, then defaults are fetched as described in Sec. 8.3.

Since this is sufficient information to build a fixed leg, it forms our definition of a fixed leg trade; see Sec. 11.5.

We can similarly build a Libor leg, again in **namespace LegBuild**:

```
───────── LegSchedule.h ─────────
Vector_<LegPeriod_> Libor
    (const Ccy_& ccy,
     const LegScheduleParams_& schedule,
     const RecPay_& rec_pay,
5    const Vector_<>& notional,
     const Vector_<>& margin = Vector_<>(),
     const Holidays_* fixing_holidays = nullptr,
     const Vector_<int>& fixing_delay = Vector_<int>(),
     const NotionalExchange_* exchange = nullptr);
```

The only subtle feature of this code is the default behavior for omitted arguments. An empty input **margin** means no margin, as does the vector [0]. But an empty input **fixing_delay** means that the fixing delay should be the default for the given **ccy**; a zero fixing delay must be explicitly specified. Omitting **fixing_holidays** likewise means that the currency default should be used.

8.5.3   *CDS*

Credit default swaps have their own conventions, separate from those of Libor swaps; in particular, their maturities are always rolled to common (usually quarterly) nominal end dates to facilitate netting.

The premium leg of a CDS (payments made by the buyer of protection) is well described as a sequence of **LegPeriod_**s, but in pricing we must include the possibility of a partial and early coupon payment due to default within a period.

For a standard CDS, the division of the protection leg into periods has no financial consequence; we need only to specify the amortization schedule (of which the protection end date is a special case). Thus a CDS trade at the time of pricing can consist of a leg of premium payments, plus a PWC_ giving the protection amount as a function of time.

8.5.4 *Inflation Instruments*

Inflation instruments introduce a different complication: the "inflation rate" on which payments are based is computed from a ratio of two price-index fixings. These indices are generally published only at one-month intervals, so an inflation leg must specify the fixing to be taken.

As a rule, we take the period start and end dates, subtract an "observation lag," and then choose whichever fixings fall in the same month as the resulting adjusted dates. This describes a "year-on-year" inflation rate; a less usual variant uses the "zero-coupon" inflation rate, where the first fixing (the denominator of the fixings ratio) is always tied to the leg's start date.

The particular date of the month on which inflation indices are fixed may be stored and accessed using the methods of Sec. 8.3.

Chapter 9

Indices

A derivative trade is a contractual agreement to exchange cashflows or other securities in the future, based on observed events in the market. To begin understanding the common features of such trades, we imagine ourselves in the far future, working out[1] the payments to which each party is contractually bound.

Through this thought experiment, we see a clear distinction between the market events – which enter the public record as *historical fixings* – and the terms of the trade, which describe the computation of a payout based on these fixings. Such historical fixings are not one-time events, but periodic (usually daily) snapshots of some ongoing process, which we call an *index*.

Indices, defined in this way, have two crucial properties. First, they stand at the interface – to a large extent, they *define* the interface – between trades and models. Once a model can simulate the dynamics of indices and of discounting, it need give no further consideration to trades.

Second, indices are a bridge uniting past and future. The computation of a payout from a set of index fixings is defined by the termsheet of a trade, and we will perform the same steps whether we are testing a scenario of the future, performing a postmortem on the past, or pricing and administering a live trade.

9.1 Naming and Parsing

An index's name plus the fixing date (or, on rare occasions, the fixing time) must completely describe its value. Thus the index name carries a lot of information, and we must work to define it with sufficient brevity.

---

[1] Amicably, of course.

The first task is to define a system of *canonical names* which provide an unambiguous representation of any index. These must be readily human-readable and also easily parsed.

We will sketch one such system, but it should not be regarded as definitive. Note that one goal is to restrict the number of special characters which must be reserved away from underlying names (*e.g.*, if _ is a special character to the parser, it is difficult to use it as part of an underlying).

We begin each canonical name with a mnemonic for the asset class, then the bracketed name of the underlying: EQ[IBM] or FX[USD/JPY]. Next we will have a separated list of increasingly-precise descriptors of the contract described by the index; as usual, our favored separator is the colon. Some indices will require an argument list (see Sec. 9.1.2); we use only named arguments and enclose the list in double-brackets. For time-shifted indices (forecasts of later values, or traded futures on an index) we use a special character (@ for a forward date, and > for a tenor or number of contracts to roll forward) immediately after the separator, followed by the date or tenor itself. Finally, fixing identifiers (nonfinancial information describing the index source, *e.g.*, identifying a particular Reuters page to use) will be appended at the end, bracketed and prefixed with FIX:. For example:

- EQ[IBM] – spot price of IBM.
- IR[USD]SWAP:LIBOR:5Y – five-year USD Libor-to-fixed swap with standard terms.
- IR[USD]SWAP:LIBOR:5Y[[FixedDayBasis=ACT/365L]] – as above but with some nonstandard terms.
- IR[USD]SWAP01:LIBOR:5Y[[FixedDayBasis=ACT/365L]] – the sensitivity of swap PV to coupon rate.
- IR[EUR]FUTURE:LIBOR@2014-09-17 – a September Euribor future.
- IR[EUR]FUTURE:LIBOR@2014-09 – a good parser will not need to be told the day.
- CM[WTI]FUTURE:>1 – the front WTI future. Note we use 1-offset.
- CM[WTI]FUTURE:>1[FIX:0#WTCL@19:30] – the front WTI future, determined by the specified published source.
- CM[WTI]FUTURE:>5bd>2 – the WTI future that will be the next-to-front contract in five business days. Note that this relies on our ability to associate WTI with NY holidays.
- CR[BCIT_SEN_DISCOUNT]CDS:5Y – a CDS rate with standard terms.
- IV[IR[USD]SWAP:LIBOR:5Y]@2014-09-17 – an implied Black vol to a fixed expiry date; see Sec. 9.4.

- `IV[IR[USD]SWAP:LIBOR:5Y]>5Y[[Normal]]` – the 5-into-5 swaption vol in normal rather than lognormal terms.

The standardized beginning of each canonical index string supports two important goals. First, it allows models to rapidly inspect indices when deciding what subset of the model is needed for pricing; see Sec. 10.2.1. Second, it allows us to replace a monolithic index parser (which is also a compile-time chokepoint) with a singleton map of asset class tags to simpler parsers. Thus the top-level parser is essentially a switch on the leading type:

```
                           IndexParse.cpp
   // does not handle composite indices; see below
   Index_* ParseSingle(const String_& name)
   {
       NOTICE(name);
5      auto stop = name.find_first_of(":[");
       if (stop == String_::npos)
           return ParseSuperShort(name);

       const String_ ac = name.substr(0, stop);
10     auto pp = TheIndexParsers().find(ac);
       REQUIRE(pp != TheIndexParsers().end(), "No parser exists");
       return (*pp->second)(name);
   }
```

This demands an index parser class with a `Parse` member, and a registration function to populate `TheIndexParsers` at load time. We will further expand this parser in Sec. 9.2.1.

9.1.1 *Short Names*

Common indices, such as Libor par swap rates, may merit a "short name" which is recognized by the parser but different from the canonical name of the resulting index. The above approach supports two ways to handle short names. Those tagged with an asset class (such as `IR:USD5Y`) will be forwarded to the parser for that asset class; while "super-short" names containing no brackets or separator (such as `USD/JPY`) have their own (hopefully simple) parser.

9.1.2 *Nonstandard Indices*

If an index differs from the expected standard, we must decide whether the trade or the model should handle the aberration. For example, if a Libor swap has an unusual coupon frequency, the trade may form an odd kind of request to the model (we show an example of this above); or the trade may synthesize the swap itself from Libor forecasts and discount factors. The former approach is preferable if models support it; but eventually we will reach the limits of such support, and require trades to make this computation themselves.

The model's capabilities should reflect those implicitly assumed by term sheets (the contracts agreeing to a trade). If a term sheet can refer to an index by describing it, as opposed to defining its calculation method from other indices, then our trades should be similarly able to describe it to the model; conversely, quantities whose calculation method is specified by a term sheet are the responsibility of the trade.

As our library matures, the range of supported indices will expand; then some trades can be simplified by rebooking them to take advantage of the new capability.

9.2   Fixings

Recall from Sec. 3.7.2 that we package access to in-process objects into the environment, and that the repository should hold the entire (unpredictable) state of our process. We use the same method to access stored historical fixings. The implementation almost exactly follows that for storage of global dates; see Sec. 5.6.4.

The available historical fixings for a given index will form a single `Storable_` object, named after the index:

```
_____ Fixings.h _____
class Fixings_ : public Storable_
{
public:
    typedef map<DateTime_, double> vals_t;
5   const vals_t vals_;

    Fixings_(const String_& index_name, vals_ = vals_t())
    :
    Storable_("Fixings", index_name) {}
10 };
```

The functionality is intrinsically simple, and there is no need to dress it up in accessors.

We provide a utility function for the common task of checking for repository access in the environment, finding the appropriate `Fixings_`, and then testing for the presence of a desired element:

```
────────── Index.cpp ──────────
double Index::PastFixing
    (_ENV, const String_& index_name,
     const DateTime_& fixing_time,
     bool quiet)
{
    static const map<DateTime_, double> EMPTY;
    auto hist = Environment::Find<FixingsAccess_>(_env);
    REQUIRE(hist || quiet, "No fixings access");
    NOTICE(index_name);
    auto fixings = hist->Fetch(index_name);
    REQUIRE(fixings || quiet, "No fixings exist");
    NOTICE(fixing_time);

    auto vals = fixings ? fixings->vals_ : EMPTY;
    auto pf = vals.find(fixing_time);
    if (pf == vals.end())
    {
        REQUIRE(quiet, "No fixing for this time");
        return -DA::INFINITY;
    }
    return pf->second;
}
```

```
────────── Index.cpp ──────────
double Index_::Fixing(_ENV, const DateTime_& fixing_time) const
{
    return Index::PastFixing(_env, Name(), fixing_time);
}
```

This does not fully address the problem of finding a fixing for an index.
The issue is the treatment of FX fixings, which can be stored in two ways;
e.g., as USD/JPY or as JPY/USD. Either of these is sufficient to define the
other. We cannot store them under the same canonical name, because they
have the same *information* but different *values* – one is about 10,000 times
larger. We do not wish to store additional information in the index, because
we want the canonical name to be a complete descriptor. Thus we must
override the Fixing method for FX indices, which therefore needs to be a
virtual method in the base Index_ class.

```
────────── IndexFx.cpp ──────────
double Index::Fx_::Fixing(_ENV, const DateTime_& time) const
{
    const double test = PastFixing(_env, XName(false), time, true);
    return test > -DA::INFINITY
        ? test
        : 1.0 / PastFixing(_env, XName(true), time);
}
```

This is not particularly efficient, but it is unlikely to be called multiple times except for complex path-dependent trades, which are not particularly rapid anyway. The utility function XName has the signature:

```
───────────── IndexFx.cpp ─────────────
String_ Index::Fx_::XName(bool invert) const
```

and of course Name calls XName(false).

This leads us to a definition of Index_:

```
───────────── Index.h ─────────────
  class Index_ : noncopyable
  {
  public:
      virtual ~Index_();
5     virtual String_ Name() const = 0;
      virtual double Fixing(_ENV, const DateTime_& fix_time) const;
  };
```

9.2.1 *Composites*

We can implement the functionality of Index_ for a linear combination of indices. This seems baroque but allows a far better description of trades such as spread range accruals, which pay a coupon proportional to the number of days in which a spread between two indices is within some contractual limits.

```
───────────── IndexComposite.h ─────────────
  namespace Index
  {
      class Composite_ : public Index_
      {
5     public:
          typedef pair<Handle_<Index_>, double> component_t;
      private:
          Vector_<component_t> components_;
          double Fixing(_ENV, const DateTime_& time) const override;
10        String_ Name() const override;
      };
```

The implementation of Fixing is obvious, while that of Name must produce a syntax we can parse. We will support addition, subtraction, and left-multiplication by a constant; unary negation is optional. Thus "*single-index - single-index*" or "*2 * single-index + 3 * single-index*" should both be legitimate.

In defining the grammar, we can require that the single index names within a composite be bracketed, thus avoiding constraints on the syntax of single indices;[2] or we can require that single index names never contain unbracketed arithmetic operators, thus ensuring that resolutions into such names (like the subtraction example above) will never break up a name. We prefer the former approach, which avoids pushing complexity down into single indices.

We now rename `Index::Parse`, from Sec. 9.1, to `ParseSingle`, and create a new function:

IndexParse.cpp

```
Index_* Index::Parse(const String_& name)
{
    if (Composite_* test = ParseComposite(name))
        return test;
    return ParseSingle(name);
}
```

Here `ParseComposite` is a local function which splits on arithmetic operators (`+`, `-` and `*`) and calls `Index::ParseSingle` as necessary. There is no provision here for nested arithmetic operations, which have no practical value.

Note that, for purposes of storing a set of fixings, we should interpret indices with `ParseSingle`, since it makes no sense to store a composite index's fixing. Also, `ParseComposite` need not consider fixing identifiers, since the arithmetic that generates the composite fixing is not subject to such specializations; accordingly, the implementation of `Index::Composite_::Fixing` should be very simple.

9.3 Sorting and Hashing

We will see later (*e.g.*, in Sec. 10.4) the need for maps with `Index_` keys. If we use `Handle_<Index_>` as the key type, we must implement:

```
bool operator<(const Handle_<Index_>& lhs,
        const Handle_<Index_>& rhs)
{   // deprecated
    return lhs->Name() < rhs->Name();
}
```

---

[2]Except that index names must not contain unmatched brackets.

to support such maps. No other implementation is readily available, since
we cannot enumerate in advance the types of indices.[3]

This is computationally expensive because **Name** must be evaluated twice
for each comparison, and many more for a binary search or map insertion.
We can do better without much implementation effort, by simply precom-
puting the index name as part of the key.

```
────────────────── Index.h ──────────────────
 struct IndexKey_
 {
     const Handle_<Index_> val_;
     const String_ name_;

5
     IndexKey_(val_)
     :
     name_(val_.Empty() ? String_() : val_->Name())
     {  }

10
     const Index_* operator->() const { return val_.get(); }
 };
 inline bool operator<
     (const IndexKey_& lhs, const IndexKey_& rhs)
15 { return lhs.name_ < rhs.name_; }
 inline bool operator==
     (const IndexKey_& lhs, const IndexKey_& rhs)
 { return lhs.name_ == rhs.name_; }
```

This can be used in place of **Handle_<Index_>** when an associative array
uses indices as keys. The same approach supports hash maps; we will hash
the precomputed **name_**, ignoring the **val_**.

9.4 Implied Vol

In some highly structured products, we are called on to observe the implied
volatility of some underlying index. This implied vol is itself an index,
containing an underlying index within itself. The examples in Sec. 9.1
show a naming scheme which makes such indices easily parseable.

When searching for historical fixings, the implied vol is treated like any
other index. During model valuation, the model must be able to compute
implied volatilities and also have knowledge of the underlying index; thus
the latter is part of our new index's public interface.

---

[3]One could imagine comparing the vtable addresses of indices, thus saving a string
comparison for indices of different derived types, but this would be depraved.

```
────────────────── IndexIv.h ──────────────────
class IndexIV_ : public Index_
{
    Handle_<Index_> underlying_;
    Cell_ expiry;
    boost::optional<double> callDelta_;    // otherwise ATM
    VolType_ volType_;
public:
    String_ Name() const;
    // help models find what is expected
    const Index_& Underlying() const { return *underlying_; }
    boost::optional<double> CallDelta(bool invert = false) const;
    DateTime_ Expiry(const DateTime_& fixing_time) const;
    const VolType_& VolType() const { return volType_; }

    IndexIv_
        (underlying_, volType_,
         const Date_& expiry_date,
         const double* call_delta = nullptr);
    IndexIv_
        (underlying_, volType_,
         const String_& expiry_tenor,
         const double* call_delta = nullptr);
};
```

The **underlying_** index should not contain a trade or fixing identifier, since we can expect to see market implied vols only for market-standard indices. The expiry might be a **DateTime_** or (more commonly) a tenor **String_**; the function **ExpiryDate** handles these two cases internally, so other functions do not need to. If a specific **call_delta** is not specified, we assume the option is ATM.[4]

Implied vol computations within a model must be reasonably rapid, but also consistent with the model's own dynamics. Suppose we construct an exotic trade where we buy an option at the market implied vol on some future date, then hold it to expiry; the modeling standard we aim for is to reliably price this trade at zero. In more complex models, we will be forced to accept some imperfect approximation to the implied vol.

---

[4]Thus ATM and 50-delta are not the same. This is an irritating quirk of the FX markets.

Chapter 10

Pricing Protocols

We have displayed an array of tools underpinning the creation of trades, forward curves, models and numerical pricers. To put these pieces to use, we have to define the target. A trade, for example, will be useful if it does certain trade-like things: but what precisely are those?

In answering this question, we are creating *protocols* for communication between high-level types. Regrettably, these protocols are usually not made explicit or given any serious design effort. There is no avoiding the existence of protocols – except by wholly sacrificing reuse – but without an active effort they will be implicit, haphazard and scattered amongst unrelated code.

This chapter is the heart of this book, and introduces many new abstractions which may be unfamiliar. Much of what has come before is original, but part is widely accepted conventional wisdom. This chapter, and those that follow and build upon it, are the result of a unique effort to analyze and systematize protocols. We introduce terminology which is new and perhaps idiosyncratic, if only because there is no other voice in the conversation – to my knowledge, no one else has explored this territory.

Readers wishing for a real-world anchor in this sea of abstraction may wish to refer occasionally to Secs. 10.13 and 10.14, which describe the way our protocol classes interact in the most common use cases. Also, Ch. 11 contains a series of example trades, beginning with the simplest cash trade, showing how each implements its share of these concepts.

As we progress through the chapter, it will gradually become clear that *protocols are binding* without the need for any explicit enforcement mechanism. Once we have established the mechanisms for communication between trades and models, the author of any new model must use them, because trades will not supply any other pricing interface.

183

10.1   The Road to Protocols

Individual parts of a derivatives library are designed by their creators, but the library as a whole generally emerges from a gradual evolution. This is largely due to the need for continuity: the business we are serving relies on uninterrupted support and demands augmentations within already-running systems. A second cause is the scale of the problem; quants are often slotted into specialized roles, and few have the wide perspective needed to design something as vast as a pricing system.

The major purpose of this volume is to offer a designed alternative to this messy accretive process. The concepts here were formed in an ongoing project to make these hidden decisions explicit, and to refine and perfect their design.

To make the following discussion more concrete, we will use a simple hypothetical case. Imagine supporting a small rates derivatives desk, whose most sophisticated products are Bermudan swaptions and (non-callable) path-dependent Libor trades; and suppose that we, as the desk's quants, have one- and two-factor Hull-White models in production and are finishing the implementation of a BGM model.[1]

Already, in this scenario, our library will embody several deeply embedded policy decisions. For instance, how are the path-dependent trades priced in both the one- and two-factor models? There are several ways to accomplish this, each with its own motivation and bringing its own set of problems.

10.1.1   *Trade-Major Pricing*

We might allow the trade to inspect the models as it sees fit, driving a Monte Carlo process appropriate to the model. This *trade-major* organization is in fact reasonably common. In its most naive form it is catastrophically clumsy, requiring separate code for each trade/model combination; but it can be improved somewhat by providing models with a polymorphic interface, with functions like `StepState` to advance the model state through one Monte Carlo time step, and `LiborRate` to convert the model state to a Libor rate the trade can use in its payout.

What should be the arguments to `StepState`? For a Hull-White or (normal) HJM model, the evolution of N state variables between two given dates can be summarized by N drift terms and an $N \times N$ covariance matrix;

---

[1] Any resemblance to the author's career should be treated as coincidental.

we could combine these[2] in a `StepParameters_` structure, and put the state itself in a simple `Vector_`. This is adequate for one class of models, but a quality implementation of a lognormal or shifted-lognormal BGM model requires stochastic drift terms; so we need a place to put those. As we consider products with different underlyings, this problem will recur repeatedly, and we will be driven to make `StepParameters_` polymorphic. As long as the model implements `StepState` this will be inconvenient, requiring close coupling between the parameters and the function implementation and forcing us to cast to the needed parameter type. This is a symptom of a design flaw, and its solution is to make `StepState` a virtual interface function of the step parametrization, which becomes a full-fledged *model stepper*.

Notice that a `ModelStepper_` is not a model or part of a model; it is *constructed by* the model for a specific task. This is an example of the principles of Sec. 2.3; by distinguishing between the original model and the steppers it creates for a pricing task, we achieve a much cleaner separation of concerns. Such a `ModelStepper_` is our first example of a *protocol class*.

Now recall that, in addition to path-dependent exotics, we also must price Bermudan swaptions. Sec. 7.10 describes how this can be made to work acceptably for a BGM model; but for a one-factor Hull-White, it would be far preferable to get the fast and precise results from a finite-difference PDE solver. Within a trade-major architecture, this gives the trade another responsibility – getting model parameters relevant to a PDE solver and driving the solver – which is obviously far from its core objective, of describing a contingent claim. This is another symptom of an erroneous policy.

10.1.2 *Path Tableaus*

Another approach to Monte Carlo pricing, reasonably common in equity exotics, is for the trade to specify a set of *event times* whose market data will drive the payout, and the market observables (*e.g.*, equity spot prices) needed at those times. The model (or its delegate) is then responsible for populating a tableau of data, in this case a three-dimensional array indexed by path, event time, and observable.[3] The trade will take these values as inputs, and compute the resulting cashflows for each path. Backward

---

[2]We would in fact store a decomposition, as described in Sec. 4.6, for efficiency's sake.
[3]Possibly we would never create the entire tableau, instead choosing to compute only one path at a time.

induction actions, such as American Monte Carlo exercise, can be handled after the payment computation.[4]

This approach is less suitable for fixed-income products. We can handle stochastic discounting of spot payments by adding an extra observable to the tableau, but forward payments require more work. Some kinds of path aggregation, *e.g.* for range accrual trades, will be unnecessarily slow. There is no obvious way to handle market observations other than equity prices; for that we need to introduce the concept of an *index*, meaning any market quantity that contains a time series, as discussed in Ch. 9. Also, when the indices used in the tableau are not spot prices, there is the chance that computing them from the model state will have a nontrivial cost. Thus the tableau, which demands computation of every index fixing on each event date, is inherently inefficient for trades with multiple legs.

Finally, the tableau does not really address the problem of PDE valuation; for that, each trade must implement its payout a second time.

10.1.3   *Protocol Design*

The effort to navigate the problems just described is the impetus for a formal set of carefully designed protocols. To maximize reuse and interoperability, we will break the pricing process into separate conceptual steps, each of which might be implemented polymorphically. Inevitably, this adds a level of abstraction to our understanding of the valuation process, and forces us to think about pricing as a generic process even when we are really pricing a simple trade with a simple model. Some otherwise capable quants have proved unable to make this step.

- Protocols must be low-level, communicating with other protocols using fundamental types. Direct use of high-level types (*e.g.*, trades, models or solvers) would defeat their purpose.
- A protocol should be designed for a single task, and should fully encapsulate the information required for its task, with any precomputation done during (or before) its construction.
- Protocols are meant to be used by higher-level pricing engines; their job is to contribute information which will be used to advance the state of a computation, not to store that state themselves.[5]

---

[4]Our final set of protocols will contain a more general backward induction action.
[5]Except for the `ValuesStore_`, which does precisely this.

- Protocol interfaces should avoid imposing unnecessary constraints on how the larger engine works. For example, a trade's payout does not manipulate the values within a PDE grid pricing to account for the possibility of early exercise; instead, it describes the necessary backward-induction actions, so that a PDE solver or American Monte Carlo can then apply them.

- The pricing process, once the participating protocol objects are created, should be as efficient as possible. Thus we create tag identifiers whose main purpose is to replace an expensive lookup or binary search with a dereference or indexing operation. Under the hood, such tags are inevitably pointers or integers; we use different C++ types for different tags to detect misuse.

- When one object must respond to complex information provided by another – most notably, when a model must provide simulated values of whatever index fixings a trade happens to require – we create a *request* object which is like a form filled out by the requester.[6]

With these ideals in mind, we turn to the problem of choosing a sufficient set of protocols.

10.2 Cast of Participants

Since our central task is pricing, we begin by considering the information flow during a *numerical pricing* task. (Throughout this work, we use the term "numerical" for computations reliant on discretized or simulated model dynamics. Closed-form or "semianalytic" pricing, which often requires a much more intimate communication between trade and model, is discussed in Ch. 14.)

Bear in mind that the model should in no way be specific to the trade: in particular, it may describe underlyings which are not relevant to the trade or may parametrize curves far beyond the trade's maturity. Thus the trade must begin by communicating its underlyings, and the model must respond by specializing to these. Recall from Sec. 2.3 that this is not accomplished by the model's changing itself, but by producing a new object of a different type.

---

[6]Such objects are passed by non-const reference, contrary to our usual practice, because their sole purpose is to be "filled out."

At this point we also commit to a numeraire currency, or *value currency*; this can be an input to the pricing request or an attribute of the trade. We prefer the latter; in the unusual event that we need to change value currencies, we can embed the trade in a composite (see Sec. 11.9).

Thus the model is responsible for three things: it must describe the joint evolution of its internal state variables and of the pricing numeraire; compute index values from the state variables at a node; and compute discount factors for whatever payments arise from the trade. For the first task, the model creates *steppers*, each of which (for a Monte Carlo) takes a step based on input random deviates and (for a PDE) computes coefficients as in Sec. 7.9. For the others, the model produces a *value request* to be sent to the trade – this is like a form for the trade to fill out to state what index values it will need to see, and what payments it may make.

The value request makes *promises* to the trade about where the needed values may be looked up in the future, by returning *tags* which will enable the lookup. The trade combines these promises to create a *payout*. In fact, the request is filled out in the process of creating the payout; this provides a very natural idiom for payout construction, where the trade simply provides information in exchange for the tags which the payout will need.

The event times (on which indices are observed) are determined from the request, and used to form the necessary steppers. The model then creates an *asset model* which honors the promises made to the trade by the request; it makes realizations of market observables available to the trade's payoff function. Similar promises are made for discount factors required for the trade's payments; these are not visible to the trade but the model handles them identically.

For backward induction (*i.e.*, PDE) pricing, this is all the model can do: the trade will manipulate values in the PDE grid, and the numerical engine will roll them back, until we reach the present.

Monte Carlo pricing allows additional control over the process, and also introduces other sources of complexity.

- The payout may be path-dependent, *i.e.*, it may have a *state* of its own.
- The model may present events – such as defaults – which are not bound to the trade's event times.
- Bermudan exercise is no longer easily handled by the payout.
- The trade may depend on the daily path of indices, as for range accrual trades.

The last of these can be avoided, if we drop the idea of holistically representing the path of an index; *e.g.*, we would think of a range accrual trade as having a trade event on every business day. However, a whole-path representation – an *index path* – is a crucial performance improvement for such trades, and is also very useful in other contexts.

To deal with Bermudan exercise decisions, the payout must describe its exercises in the form of *backward induction actions* which can be taken by the numerical pricer after all paths have been run.

The backward induction mechanism is also often used for barrier events. These could be treated in the trade's state during forward induction, but doing so would introduce discontinuities in the payout which would defeat any attempt to compute a numerical hedge ratio. Instead, during forward induction we can compute payments for both trade states, then mix them using a smoothed hit-probability function. This enables stable risk computations, and allows barrier shifts to be supplied at the time of pricing rather than embedded in the trade.

10.2.1 *Which is a Model?*

Even in the heuristic discussion above, we have risked confusion by using the word "model" in conflicting ways. There are two distinct roles here. The first is an object that defines a function mapping trades to values; this would properly be called a "pricer", but "model" is the established nomenclature. The "model" may lack any consistent dynamics (as in a swap market model) or may recalibrate internal parameters in a trade-dependent way; but it is invariably still called a model, however little it may deserve the appellation.

The second use is an object that defines the evolution through time of market observables (or of some internal state which drives the observables). This is a model in the mathematical sense; but that name is already taken. "Pricing model" sounds too much like pricers (*i.e.*, models) to be safely used. In this work we use "SDE", which is not strictly accurate but does convey the sense of a process (or set of processes) in a particular measure. Thus the model creates an SDE based on the trade's underlying and value currency; and the SDE in turn creates the steppers, value request, and asset model. We will still sometimes refer loosely to "the model" when there is no prospect for confusion, but it should be understood that it is the SDE which interacts directly with the numerical pricing engines.

By separating the model and the SDE, we also allow an important optimization; the SDE used for pricing a particular trade can omit unneeded underlyings, and truncate term structures (*e.g.*, for BGM models) after the trade's maturity. Thus the SDE is optimized for a pricing task, while the model is maximally general.

10.3   Past and Future

The above discussion is also complicated by the necessity to distinguish somewhere between "past" and "future" – between events that are now historical and those that are still uncertain. We distinguish several ways of making this distinction, each defined by its own "present".

- *Reset time* – after which index fixings after unknown.
- *Accounting date* – before which payments have "rolled off" and are ignored.
- *Value date* – on which the discount factor is 1.
- *Vol start time* – at which stochastic processes "turn on."

As a rule, all these will be the same; however, when computing various decay scenarios we may change them in any order.

The vol start time is an attribute of the model; the reset time and accounting date are parameters of the valuation, or in the environment. We will access the latter as described in Ch. 5. The value date can be treated either way; on balance we find it better to treat it as a parameter of the valuation.

The reset time is the most fundamental of these, since it determines the meaning of a request for an index fixing; it partitions the historical past from the simulated future in all our numerical methods.

It is important to construct the valuation process so that the trade does not see this partition: this avoids substantial code duplication in each concrete trade type. The model will be less protected, because its definition of "present" (the vol start time) may not coincide with the reset time.

We will mediate the trade's requests to ensure that the model is not responsible for lookup of historical data; see Sec. 10.16.6. For requests between reset time and vol start time, the model might give a deterministic answer; but it still must return this answer in the form of a promise.

10.4 Underlyings

The underlying is meant to support the creation of an appropriate SDE, so it provides fairly restricted information.[7]

```
─────────────────────── Underlying.h ───────────────────
struct Underlying_
{
    map<Ccy_, Date_> payCcys_;
    map<IndexKey_, DateTime_> indices_;
    map<String_, Date_> credits_;

    Underlying_& operator+=(const Underlying_& more);
};
```

For each currency paid, we note the last pay date; for each index, the last event time at which it is observed; and for each reference credit, the last default date on which it need be monitored.

An alternative is to have the underlying provide complete information about the usage times of every index, which could then be used in place of the value request. However, this is simply too much information for many purposes; thus we prefer the simpler underlying.

10.5 Payments and Streams

So far we have not discussed how the cashflows generated by the trade are to be discounted. We generally divide this discounting into two portions: a factor $D(t)$ from today to the event time, and a zero-coupon bond price $P(t, T)$ at the event time. The trade could extract the latter itself, since both $P(t, T)$ and any necessary FX rate can be represented as index fixings and extracted through the value request; but this would be a mistake.

The problem is that the discounting is absorbed into the opaque workings of the trade, and commingled with the computation of amounts. Part of our job is to support the ongoing maintenance of real trades, not just valuation; and a report of the payments to be made is an important part of this maintenance.

Thus, rather than discount its own payments, we will force a trade to register a possible payment, obtaining for each a *payment tag* which will mediate the discounting during valuation. At first glance this might seem

───────────────────────

[7]We may extend the underlying slightly to support trade-specific calibration: see Sec. 14.7.

like sheer obfuscation; but it lets us use the same machinery for valuation and for other kinds of reporting (*e.g.*, a list of expected upcoming payments). Thus it saves us from duplicative and error-prone recoding in each concrete trade type.

A payment tag is not much to look at:

```
                            ─── Payment.h ───
namespace Payment
{
  class Tag_ : noncopyable
  {
  public:
    virtual ~Tag_();
  };
  const Handle_<Tag_>& Null();

  namespace Amount
  {
    class Tag_ : noncopyable
    {
    public:
      virtual ~Tag_();
    };
  }
}
```

Null here is the equivalent of the Unix /dev/null – it is a tag for payments that do not matter (generally because they precede the accounting date). Making it visible in this way allows this fact to be communicated to the trade, allowing some unimportant optimizations.

We make a distinction in code between payments, which have financial value and may be discounted, converted to different currencies, and so on; and raw amounts, which are just numbers. This lets us write clearer and safer code with no extra run-time cost.

10.5.1 *Payment Reporting*

More interesting than the tag itself is the information which the trade provides in order to obtain it. This (also in namespace Payment) gives some description of the payment as well as the minimal information to compute the discount factor. The descriptive information is ignored during valuation but makes the payment report more useful.

This information is not discarded when the tag is formed; it can be stored by the task (*e.g.*, the valuation or reporting engine). The tag is

generally also an index key which can be used to access whatever portion
of this information is retained, plus derived information like precomputed
coefficients for discounting.

```
                          ──────── Payment.h ────────
     namespace Payment
     {
         struct Conditions_
         {
 5           enum class Exercise_ : char
             {
                 UNCONDITIONAL,
                 ON_EXERCISE,
                 ON_BARRIER_HIT,
10               ON_CONTINUATION
             } exerciseCondition_;
             enum class Credit_ : char
             {
                 RISKLESS,
15               ON_SURVIVAL,
                 ON_DEFAULT
             } creditCondition_;
             // if paid on default, we need still more info:
             DefaultPeriod_ defaultPeriod_;

20           Conditions_();   // unconditional, riskless case
         };

         struct Info_
25       {
             String_ description_;
             DateTime_ knownTime_;
             Conditions_ conditions_;
             boost::optional<AccrualPeriod_> period_;
30           Info_(const String_& des = String_(),
                   const DateTime_& known = DateTime::Minimum(),
                   const Conditions_& cond = Conditions_(),
                   const AccrualPeriod_* accrual = nullptr);
         };
35   }

     struct Payment_
     {
         DateTime_ eventTime_;
40       Ccy_ ccy_;
         Date_ date_;
         String_ stream_;
         Payment::Info_ tag_;
         Date_ commitDate_;
45       Payment_();   // support Vector_<Payment_>
         Payment_(const Payment_& src);
```

```
    Payment_
        (const DateTime_& et,
         const Ccy_& ccy,
50       const Date_& dt,
         const String_& s,
         const Payment::Info_& tag,
         const Date_& cd = Date::Minimum());
};
```

The struct Conditions_ is an indication of the kind of information that might be included in a payment report. We nest Info_ inside namespace Payment, rather than inside struct Payment_, so that it can be forward-declared in other headers.

10.5.2 *Commitment to Streams*

This introduces the concept of a *stream*, into which payments are directed. For those familiar with PDE pricing, a stream is essentially the thing whose value is rolled back on the PDE grid: it is a collection of payments which can be summed for purposes of valuation. The two legs of an ordinary swap can be collected into a single stream. A Bermudan swaption has two streams (the swap, and the option to enter into the swap) though we might in practice roll back only the latter and price the former *ab initio* at each node. In either case, the trade value is obtained from the stream values by some linear relation, which the trade must specify.

Streams are also the underlyings of options. A bond, for instance, is a stream; if we take possession of the bond at a specified *delivery date* – due to an option exercise, for instance – we receive the coupons whose payment dates are after delivery. This is represented by the commitDate_ above, which for a bond coupon is just the coupon payment date; it is the date which is compared to the delivery date to determine whether the payment is received. We will represent the accrued interest in a bond option as a fee paid upon exercise. Thus the specification of commitDate_ and of the appropriate fees let us accurately represent several different options on the same stream; Sec. 10.9 describes this in detail.

There is one complication: truly American options, or Bermudan callable swaps with an exercise frequency higher than the coupon frequency. For the latter, upon exercise we must pay accrued interest for the elapsed part of a (possibly floating) period. We can deal with this by complicating the framework, extending the commitDate_ above to an interval (start date, end date, and day basis) which will support fractionally-committed payments like this one; or by complicating the trade, breaking the coupon payments into several payments with different commit dates (but otherwise identical). The decision depends on the business mix we are expecting; here we will continue with the simpler framework.

10.5.3 *Destinations*

The precise form of the payout depends on the method for committing payments to streams. Recall from Sec. 10.1.3 that the payouts will not have direct access to the storage of values; they must operate identically on a Monte Carlo path or at a node in a finite-difference grid.

For this we use a simple abstract class, with a payment represented by a call to `operator+=`:

```
                        ──────── Payment.h ────────
    class NodeValue_ : noncopyable
    {
    public:
       virtual ~NodeValue_();
 5     virtual void operator+=(double amount) = 0;
       // direct support for backward induction:
       // virtual double& operator*() = 0;
    };

10  class NodeValues_ : noncopyable
    {
    public:
       virtual ~NodeValues_();
       virtual NodeValue_& operator[]
15        (const Payment::Tag_& tag) = 0;
       inline NodeValue_& operator[]
          (const Handle_<Payment::Tag_>& tag)
       { return operator[](*tag); }

20     virtual double& operator[]
          (const Payment::Amount::Tag_& tag) = 0;
       inline double& operator[]
          (const Handle_<Payment::Amount::Tag_>& tag)
       { return operator[](*tag); }
25  };
```

The `operator*` in `NodeValue_` requires some explanation. In Monte Carlo valuation, we simply add payments to the streams; manipulation of stream values (for example, to reflect Bermudan exercise) is not meaningful within a single path and requires separate machinery (see Sec. 10.9 on backward induction actions, and Sec. 7.10). In PDE valuation, we have two main alternatives: we can allow the trade to directly manipulate the stream values (as `NodeValue_::operator*` does), or we can require the PDE to apply the same backward induction actions formed for the Monte Carlo.

The decision whether to supply this function will be based on our opinion of the relative importance of PDE and Monte Carlo pricing. If we provide `operator*`, we can simplify the PDE implementation; but a Monte Carlo engine cannot support it, and cannot price any trade which uses it. My own preference is to use PDE pricing very little, instead favoring large, flexible models whose high dimensionality necessitates Monte Carlo pricing. In this work, I will not show PDE-specific code.

The `NodeValues_` also allows storage of amounts which are not payments, which will not be discounted or accumulated. These are used, *e.g.*, to store the observables for an American Monte Carlo which will run after the loop over paths is complete.

10.6   Index Paths

We have discussed the desirability of accessing the whole path of an index, not just discrete fixings, in computing the payout. This is a major commitment for models – once we take this route, models whose SDE's cannot produce index paths will fail to price trades which attempt to use them.

```
_____ IndexPath.h _____
class IndexPath_ : noncopyable
{
public:
    virtual ~IndexPath_();

5
    virtual double Expectation
        (const DateTime_& fixing_time,
         const pair<double, double>& collar)
    const = 0;

10
    virtual double FixInRangeProb
        (const DateTime_& fixing_time,
         const pair<double, double>& range,
         double ramp_width = 0.0)
15  const = 0;

    virtual double AllInRangeProb
        (const DateTime_& from,
         const DateTime_& to,
```

```
20         const pair<double, double>& range,
           double monitoring_interval,
           double ramp_sigma = 0.0)
       const = 0;

25     virtual double Extremum
           (bool maximum,
           const DateTime_& from,
           const DateTime_& to,
           double monitoring_interval,
30         const pair<double, double>& collar)
       const;    // default calls AllInRangeProb
   };
```

The output of each function is an expectation conditional on all state information at event times; thus use of the index path is essentially path integration over the "blank space" between event times. It follows that a linear combination of these, such as a number of days in range, is also a conditional expectation.

Our convention is that payments should be linear in the outputs of the IndexPath_ member functions. This lets us mix *a single piece* of whole-path information into the payout with complete freedom; and conditionally independent quantities (*e.g.*, days in range over disjoint intervals separated by an event time) can also be combined freely. Conversely, nonlinear functions of whole-path information (*e.g.*, a product of two extrema) would not be priced correctly by this method; in the unlikely event that one should become important, we would have to introduce a new member in IndexPath_ to evaluate it.

The implementation of the path will likely use Brownian-bridge approximations, which we work to make accurate but which cannot be perfect. The errors thus introduced can be measured, and if necessary controlled, by inserting fake event times into the payout – this will more tightly constrain any Brownian interpolation and approach the limit of using only information at event times.

10.6.1 *Historical Paths*

A path which is completely in the past – *i.e.*, all its fixings are before the reset time – clearly should have a model-independent implementation. Also, such a path might need to be part of a model-constructed IndexPath_; *e.g.*, the times in a call to Extremum might span the reset time. We create a transparent data structure for this purpose:

```
  struct IndexPathHistorical_ : IndexPath_
  {
      map<DateTime_, double> fixings_;
      double Expectation
5         (const DateTime_& t,
           const pair<double, double>& lh)
      const override;
      // ...
```

The implementations to complete the `IndexPath_` interface are uniformly simple. In the unusual case of trades that depend on intraday prices, we may need to extend `IndexPathHistorical_` to store open-high-low-close information.

10.7   Defaults and Contingent Payments

We may need to price trades which depend on the defaults in some pool of reference securities; thus the presence of `Underlying_::credits_` above. Should the model simulate a default event, this must be communicated to the trade and to the numerical method, which are concerned with different aspects of the event. Thus we first isolate one part for the trade to see:

```
──────────────── Default.h ────────────────
  struct ObservedDefault_
  {
      Date_ date_;
      CreditId_ referenceName_;
5     double recovery_;
  };
```

Then we write, in another header not seen by trades,

```
──────────────── DefaultModel.h ────────────────
  class DefaultEvent_ : noncopyable
  {
  protected:
      DefaultEvent_
5         (const Date_& date,
           const CreditId_& which,
           double recovery,
           double df_from_event);
  public:
10    virtual ~DefaultEvent_();

      const ObservedDefault_ observed_;
```

```
        const double dfFromPreviousEvent_;
};
```

Subclasses of `DefaultEvent_` may contain other information seen only by the SDE and its steppers; the trade and Monte Carlo will not care. The numerical pricer will show only the `observed_` part of the default event to the trade, and will use the discount factor to discount payments made by the trade in response to the default.

The `CreditId_` is a proxy for the reference name of the defaulted credit. Multi-credit trades will almost surely need to look up this identifier in a sizeable table of names; *e.g.*, a map of names to notionals for a CDO. We introduce `CreditId_` to avoid making a large number of string comparisons, which can add significantly to the running time. Its role is simply to support rapid comparison:

```
———————————————————— Default.h ————————————————————
   struct CreditId_
   {
       int val_;
       explicit CreditId_(val_) {}
 5 };
   inline bool operator<(const CreditId_& lhs,
           const CreditId_& rhs) {return lhs.val_ < rhs.val_;}
```

The id's must remain valid for the duration of the pricing. In practice, we accomplish this by storing the reference credits in an object that will not be destroyed before pricing is complete, then doling out their addresses within that object.

```
———————————————————— Default.h ————————————————————
   struct AssignCreditId_ : noncopyable
   {
       Vector_<String_> names_;
       AssignCreditId_(const Vector_<String_>& names);
 5     CreditId_ operator()(const String_& name);
   };

   CreditId_ AssignCreditId_::operator()(const String_& nm)
   {
10     auto pn = LowerBound(names_, nm);
       REQUIRE0(pn != names_.end(), "Reference credit '" + nm
           + "' is not part of the trade underlying");
       return CreditId_(pn - names_.begin());
   }
```

The constructor of `AssignCreditId_` will sort the input names for faster lookup.

The pricing routine will create a single instance of this object, based on the trade's stated underlyings, to be shared by the trade and SDE during the setup phase. Its mission is then accomplished: it has remained in scope long enough to ensure that each unique credit has been assigned a distinct identifier.

10.7.1 *Immediate Payments*

Ideally, the trade (*i.e.*, the payout) would receive the default information, update its state accordingly, and then make payments at the next event time. However, a trade might generate cashflows immediately upon default, rather than politely waiting for an event time, and we must prepare for that contingency. Thus we need a destination, akin to `NodeValues_`, for such immediate payments:

```
                        ─── Payment.h ───
class NodeValuesDefault_ : noncopyable
{
public:
    virtual ~NodeValuesDefault_();
5   virtual NodeValue_& operator()
        (const Payment::Default::Tag_& tag,
         const Date_& commit_date) = 0;
    inline NodeValue_& operator()
        (const Handle_<Payment::Default::Tag_>& tag,
10       const Date_& commit_date)
    {
        return operator()(*tag, commit_date);
    }
};
```

Here the tag (now of type `Payment::Default::Tag_`, but just as blank and uninteresting as `Payment::Tag_`) holds information about the stream into which the payment is made. Since the commitment date (see Sec. 10.5.2) cannot be determined without knowing the default date, it is left for the trade to supply along the path.

10.7.2 *Viewing Indices*

The payment might not be truly immediate: more commonly, it takes place some days or months after the default event. Who should be responsible for discounting the payment? While we usually avoid making trades responsible for any discounting computation, it turns out that discounting from the default date to the payment date is much more simply done by the payout: the trade can describe the necessary discount factor as an `Index_`, and the payout can read its value from the `IndexPath_`.

This shows the desirability of making index paths visible to the payout when a default occurs; also, the paid amount might itself depend on other market quantities at the time of default.

10.8 Requests and Promises

The form of the value request is dictated by the form of the promise it makes. This is some abstraction of an address: a location to which the needed value will later be written. Indeed, it can be a literal address, probably `volatile const double*` for values – this maximizes efficiency, while a less explicit promise will leave us more implementation freedom. In practice, the literal address makes multi-threading unnecessarily complicated, and we will avoid it. We would prefer to promise something compact and inscrutable, like an `int`, and delegate its interpretation completely to the asset model.

```
──────────────── AssetValue.h ────────────────
namespace Valuation
{
    typedef size_t address_t;
    boost::optional<double> KnownValue(address_t loc);
5   boost::optional<address_t> FixedLoc(double value);

    struct IndexAddress_
    {
        int val_;
10      IndexAddress_(val_) {}
    };
}
```

But we can hide rather a lot inside a (32-bit) integer! One useful thing to hide is an indication of time or event index, which can be used inside the asset model to ensure that the trade is not "peeking ahead" to part of the path which has not yet been formed. This is not part of the interface to trades, which receive an `address_t` already formed and later use it, unaltered, to obtain a value from the model. A trade and its payout have no idea what an `address_t` contains or points to; they simply use it in a prescribed way to obtain the market observations they require.

We can also hide a set of fixed values inside an address: the trade can use `KnownValue` to see if the address points to a deterministic quantity, and the SDE can use `FixedLoc` to see whether a given number has a value. (The latter function can contain a singleton registry of addresses and add to it as needed, but this is hostile to multi-threading.) This can sometimes support optimizations within the payout, though our preference is to use it only for `0.0` and `1.0`.

Index path addresses have their own type, and here `typedef` is not a strong enough distinction: two typedefs which resolve to the same type are interchangeable to the compiler. Thus we create a separate `struct` for this address type, though in all probability it also wraps an `int`.

If there is no abstraction penalty in performance, we can do the same for value requests, replacing `address_t` with a `struct`. The best test for abstraction penalty is to change the code and see whether the binary (non-debug) dll's change at all. Essentially any change represents a performance degradation.

The asset model's contents in their entirety will not be available to the payout. Instead we will send a *token*:

```
_____ AssetValue.h _____
   class UpdateToken_
   {
       typedef Vector_<Handle_<IndexPath_>> indices_t;
       Vector_<>::const_iterator begin_;
5      indices_t::const_iterator indexBegin_;
       const int valMask_, dateMask_;
   public:
       const DateTime_ eventTime_;
       UpdateToken_
10         (Vector_<>::const_iterator begin,
            indices_t::const_iterator index_begin,
            int val_mask,
            int date_mask,
            const DateTime_& event_time);
15
       inline const double& operator[]
         (const Valuation::address_t& loc) const
       {
           return *(begin_ + (loc & valMask_));
20     }
       const IndexPath_& Index(const Valuation::IndexAddress_&) const;
   };
```

By design, the token is the narrowest possible window into the state of the asset model. In a better world, `UpdateToken_` would be an abstract class with a virtual `operator[]`; but here, at the hottest of all code hotspots, we begrudge even the overhead of an arithmetic operation, and cannot tolerate a virtual function call. Thus the implementation commits to a convention for the use of the bits in an `address_t` – namely, that the least significant bits will describe an offset into a vector of market observations. The remaining bits can be used to aid debugging, *e.g.* by `asserting` some expected relation between `loc&dateMask_` and the `eventTime_`.[8]

The `eventTime_` must be part of the token's public interface, so that payouts can look at the token to determine what actions to take.

10.8.1 *The Value Request*

Now, who will make these promises to the trade?

```
──────────────── ValueRequest.h ────────────────
class ValueRequest_
{
public:
    virtual ~ValueRequest_();

    virtual Handle_<Payment::Tag_> PayDst(const Payment_& p) = 0;

    virtual Handle_<Payment::Default::Tag_> DefaultDst
        (const String_& stream) = 0;

    virtual Valuation::address_t Fixing
        (const DateTime_& event_time,
        const Index_& index) = 0;

    virtual Valuation::IndexAddress_ IndexPath
        (const DateTime_& last_event_time,
        const Index_& index) = 0;

    typedef Valuation::address_t address_t;
    typedef Valuation::IndexAddress_ IndexAddress_;
};
```

We have already described the `ValueRequest_` as a form to be filled out by the trade, in the process of creating a payout. To ensure the form is filled out completely, we allow it to hold hostage the `Tag`s and `addresses` which the payout will require, yielding them up only when given sufficient information for the SDE and numerical method to populate them. Thus the request makes no promises that cannot be kept; and the successful creation of a payout ensures that it can be priced. This scheme makes the

---

[8]We lack enough bits to enforce equality.

marshaling of information very explicit, sometimes painfully so; but with it, we experience very few run-time errors.

The payout, once created, will be able to compute the cashflows at each node, as soon as it receives the **UpdateToken_**. The token simultaneously grants access to the values stored at the promised addresses, and testifies that those values are up-to-date and ready for use – *i.e.*, that the stepper and model asset have done their jobs.

10.8.2 *Help for Models*

In a separate header, not visible to trades, we expand this interface slightly:

```
                        ── ValueModel.h ──
   class IndexPathHistory_ : noncopyable
   {
   public:
       virtual ~IndexPathHistory_();
5      virtual DateTime_ ResetTime() const = 0;
       virtual Handle_<IndexPathHistorical_> History
           (const Index_& index)
           const = 0;
   };
10
   class ValueRequestImp_ : public ValueRequest_
   {
   public:
       virtual address_t InsertFixing(double amount) = 0;
15
       virtual IndexAddress_ IndexPath
           (const DateTime_& last_event_time,
            const Index_& index,
            const IndexPathHistory_* historical) = 0;
20
       IndexAddress_ IndexPath
           (const DateTime_& last_event_time,
            const Index_& index)
           {
25         return IndexPath(last_event_time, index, 0);
           }

       virtual IndexAddress_ InsertPath
           (const Handle_<IndexPath_>& path) = 0;
30
       virtual map<IndexKey_, address_t> AtTime
           (const DateTime_& t) const = 0;
   };
```

This allows the SDE to see what has been requested, so that it can produce the necessary updaters.

As discussed in Sec. 10.3, the `ValueRequest_` must distinguish between requests before and after the reset time, so that the trade will not need to. In practice, the SDE will form a `ValueRequestImp_` object for its own use;[9] the numerical method will then decorate it (see Sec. 2.6), wrapping it in a different `ValueRequestImp_` which handles requests in the past before the SDE sees them. It is the latter wrapper which will be seen by the trade. The `InsertFixing` and `InsertPath` functions support this separation; see Sec. 10.16.6.

The `AtTime` function allows the `SDE_` to discover what promises have been made by a `ValueRequest_`; see Sec. 13.3.

10.8.3 *Destinations*

In PDE valuation, we will be accumulating a separate value for each stream (thus our PDE solver, in Sec. 7.9, handles a vector of payouts); the output of `PayDst` is an abstraction of the index into this vector. In Monte Carlo valuation, where accumulation often must be deferred, it is a more detailed label. The trade specifies these destinations by providing the `stream_` containing the payment (see Sec. 10.5.1), but it knows nothing of the accumulation mechanism; it simply treats the tag as a kind of subscript.

Payments before the accounting date can be ignored; we accomplish this by returning `Payment::Null()` from the `ValueRequest_`. The trade might check the returned handle (by pointer equality) and optimize accordingly (by not computing the amount at all). In practice, since a payment in the past can only be made by an event also in the past, its amount will not be computed more than once; thus this is not a significant optimization.

10.9 Bermudans and Barriers

A European option can be priced using Monte Carlo in two ways: as a cap or floor (often zero) on the payment amount into a single stream, or as a conditional transition from one stream (in which no payment is made) to another (where the option's underlying is paid). The former approach is simpler, especially for trades containing several independent options (*e.g.*, interest rate caps or cliquets). The latter is necessary, however, for

---

[9]It turns out that all SDEs can share a single implementation of `ValueRequestImp_`.

Bermudan exercises when there are several opportunities to make the transition between streams (or even for Europeans when the underlying itself requires Monte Carlo valuation).

Before we can define an option, we must be able to define what is received upon exercise. This is a combination of single payments (which we refer to as *fees*, regardless of their sign) and portions of streams. First we define these portions:

```
─────────────────── BackwardInduction.h ───────────────────
   namespace BackwardInduction
   {
       using Payment::Tag_;    // for fees
       typedef Payment::Amount::Tag_ amount_t;
 5
       struct StreamSegment_
       {
           String_ stream_;
           Date_ deliveryDate_, terminationDate_;
10         StreamSegment_
               (const String_& s,
                const Date_& delivery,
                const Date_& termination = Date::Maximum());
           StreamSegment_();    // support vectors
15     };
```

A combination of fees and **StreamSegments** is adequate to describe any option's underlying. Given this representation, we form different concrete actions to take during backward induction.

To define the result of a Bermudan exercise decision, we must know not only the stream and delivery date but also the sign (long or short) of the option, the fees paid and streams received on exercise (if any), and the observables (see Sec. 7.10).

```
─────────────────── BackwardInduction.h ───────────────────
       struct Exercise_
       {
           int sign_;
           // where to find the values
 5         Vector_<Handle_<Tag_>> fees_;
           Vector_<StreamSegment_> underlyings_;
           Vector_<Handle_<amount_t>> observables_;
           Vector_<Handle_<amount_t>> slaves_;    // write exProb
       };
```

We also prefer to think of barrier hitting, in a knock-in or knock-out option, as a stream transition. It is equally possible to think of it as a modification to the payout state, which then affects all later payments; but making the stream transitions explicit has substantial operational value. There may be payments if the barrier is hit, or streams which knock in; we would like to report the probability of hitting; and, as mentioned at the start of this chapter, postponing the decision of whether the barrier was hit until the backward induction phase allows us to stabilize the computed risk.

```
────────────────── BackwardInduction.h ──────────────────
      struct Barrier_
      {
          Vector_<Handle_<Tag_>> payOnHit_;
          Vector_<StreamSegment_> knockIn_;
          Handle_<amount_t> hitProb_;
      };
```

Handling barriers in the backward induction phase is also important for some applications, such as asset swaps on callable bonds, where it is necessary to make a change to one stream when a different stream is called. The former stream is said to be "slaved" to the latter. This is the purpose of `Exercise_::slaves_`; we create a `Barrier_` action in the slaved stream, and add its `hitProb_` member to the list of `slaves_` of the exercise point in the master stream. Upon making the exercise decision, we write the result (0 or 1) to that location; it is then available for use by the slaved stream.

The `HitProb_` is an amount, not a boolean flag, for several reasons. First, we already have a mechanism for transmitting values. Second, the value may be determined from an index path using `FixInRangeProb` or `AllInRangeProb`, which can give values over the whole range $[0, 1]$. Finally, we will often artificially smooth the hit probability to give continuous and/or conservative values; continuous values are crucial for numerical hedge computation.

Another action, `Include_`, simply adds one stream's payments to another. This lets us write trades, should we choose, with finer granularity: we can represent legs as separate streams, then pull them together using `Include_`.

```
────────────────── BackwardInduction.h ──────────────────
      struct Include_
      {
          vector<StreamSegment_> src_;
      };
```

Exactly one of the above specific actions is required to form a valid generic action. The proper high-tech solution is a `boost::variant`, which

accomplishes exactly this. Since we must support vectors of actions, we
need a default constructor which perforce contains no valid action; thus we
add boost::empty as an alternative action type.

```
───────────── BackwardInduction.h ─────────────
     struct Action_
     {
         String_ stream_;
         DateTime_ eventTime_;
5        Date_ deliveryDate_;
         variant<Exercise_, Barrier_, Include_, Empty_> details_;
         Action_(stream_, eventTime_, deliveryDate_);
         Action_();     // support Vector_<Action_>
     };
10 }    // leave namespace BackwardInduction
```

The Action_ is a passive struct, communicating data from the trade for
use by the numerical pricing engines.

10.10 Payouts

At last we can define the payout of a trade.

```
───────────── Payout.h ─────────────
     class Payout_ : noncopyable
     {
     public:
         typedef Handle_<Payment::Tag> dst_t;
5        typedef Handle_<Payment::Amount::Tag> amount_t;
         virtual ~Payout_();

         virtual Vector_<DateTime_> EventTimes() const = 0;

10       class State_ : noncopyable
         {
         public:
             virtual ~State_();
             virtual State_* Clone() const = 0;
15           virtual State_& operator=(const State_& rhs) = 0;
         };
         virtual State_* NewState() const { return nullptr; }
         virtual void StartPath(State_* state) const {}

20       virtual void DoNode
             (const UpdateToken_& values,
              State_* state,
              NodeValues_& pay_dst)
         const = 0;
25
```

```
       virtual void DoDefault
         (const ObservedDefault_& event,
          State_* state,
          const NodeValuesDefault_& pay_dst)
30     const
       {   }   // default implementation is no-op

       virtual Vector_<BackwardInduction::Action_> BackwardSteps()
       const = 0;
35
       // components of value for each trade name
       typedef map<String_, Vector_<pair<String_, double>>> weights_t;
       virtual weights_t StreamWeights() const = 0;
};
```

The final member function, **StreamWeights**, provides the translation from stream values to trade values. It takes this complex form because a trade may have several payouts (Sec. 11.9 describes composite trades; we can even have customized trades involving more than one counterparty), each of which is a linear combination of stream values. The top-level trade names are the keys of the output of **StreamWeights**.

10.10.1 *Trade State*

Payout_::State_ is simply a placeholder to be carried by the Monte Carlo; each path-dependent trade will create its own subclass using **NewState**, which will be returned to it at each call of **DoNode** or **DoDefault**. A PDE, which naturally does not support path dependence, will pass **nullptr** instead. Thus payouts must check that the state exists before using it, unless we add some other way to signal path dependence. **StartPath** will be called by the Monte Carlo pricer at the start of each path, so the **State_** can be reused without fully re-creating it.

The base implementations of **NewState** and **StartPath** are those appropriate for stateless (non-path-dependent) trades.

10.10.2 *Values Store*

The **NodeValues_** must translate from *amounts* of the cashflows made by the trade to *values* which can be discounted, converted to a common currency, and summed over payment times. To accomplish this, we will require market information during the construction of the **NodeValues_**; we provide the same **ValueRequest_** used in constructing the **Payout_**, so that these requests can be joined with the trade's direct market observations.

10.11   Steps

In contrast to the machinery above, the interface between SDEs and numerical methods is quite simple. Partly this is due to the extreme opacity of models: they have state variables, which change stochastically through time; they use these state variables to generate observable fixings; and there ends the story.[10]

For the PDE, the interface of our solver (Sec. 7.9) determines what information the stepper must provide. The Monte Carlo interface is more opaque – since the dynamics are completely concealed – but richer because of additional possibilities:

- The steps may be path-dependent (*e.g.*, a credit step may depend on which credits have already defaulted), and this information should be communicated to the stepper.
- The steps may store a *paths record* with information about the run, which can be used to stabilize risk in bumped runs.
- The stepper may be shared by multiple threads; thus it must be fully bitwise constant (with no `mutable` or `shared_ptr` members).

We require a few supporting classes first:

```
────────────────────── MCPath.h ──────────────────────
     namespace MonteCarlo
     {
         // memory used by the stepper
         struct Workspace_ : noncopyable
 5       {
             virtual ~Workspace_();
             virtual Workspace_* Clone() const = 0;
         };
         struct PathsRecord_ : noncopyable
10       {
             virtual ~PathsRecord_();
             virtual void StartPath(int i_path) = 0;
         };
         typedef shared_ptr<PathsRecord_> record_t;
15   }
```

A `PathsRecord_` spans the whole duration of a path, unlike a stepper which is responsible only for a single time step; the various steppers will all see the

---

[10]Our use of a `Vector_<>`, rather than a generic `struct`, to store model state is a concession to the PDE.

same instance of `PathsRecord_`. Since its main use is to store information needed to stabilize a later bumped run, it is also reused across paths (a viable alternative would be to have a separate `PathRecord_` for each path).

Each individual stepper creates whatever `Workspace_` it needs;[11] the `record_t` is created by the parent SDE, and made available to the steppers to be shared with their workspaces.

Workspaces are not shared across threads (we `Clone` them instead), so they are the appropriate place for any working memory the stepper may require.

```
                            ───── Step.h ─────
     class ModelStepper_ : noncopyable
     {
     public:
         virtual ~ModelStepper_();

 5
         // PDE interface
         virtual PDE::ScalarCoeff_* DiscountCoeff() const;
         virtual PDE::VectorCoeff_* AdvectionCoeff() const;
         virtual PDE::MatrixCoeff_* DiffusionCoeff() const;

10
         // MC interface
         virtual MonteCarlo::Workspace_* NewWorkspace
             (const MonteCarlo::record_t& paths_record)
         const = 0;
15       virtual int NumGaussians() const = 0;
         virtual void Step
             (Vector_<>::const_iterator iid_gaussian_begin,
             Vector_<>* state,
             MonteCarlo::Workspace_* work,
20           Random_* more_randoms,
             double* rolling_df,
             Vector_<Handle_<DefaultEvent_>>* defaults)
         const = 0;
     };
```

In addition to sharing the paths record across steps, for many models the parameters used in a stepper depend on some quantity accumulated before the step start date. For example:

- In BGM-like models, we may linearize the drift, which can be done more accurately if we know the moments of the state at the step start time.

---

[11]Bear in mind that different steppers for the same SDE may have different sizes, *e.g.*, as FRAs roll into the past in a BGM model.

- In Markov chain models, jump compensator terms depend on the distribution of the Markovian state.

These and similar computations can in principle be performed *ab initio* for each stepper being created, but – especially in the presence of many short steps – it is far more efficient to maintain and update a cumulative state. For this to work, the Monte Carlo driver must create the steps in chronological order.

Since we will use this object in the creation of steppers, it makes sense to associate the paths record with it as well:

```
                          ─────── Step.h ───────
    struct StepAccumulator_ : noncopyable
    {
        virtual ~StepAccumulator_();

5       virtual MonteCarlo::PathsRecord_* NewPathsRecord
            (int num_paths,
             const MonteCarlo::record_t& base_record)
        const
        { return 0; }
10
        virtual Vector_<pair<double, double>> Envelope
            (const DateTime_& t,
             double num_sigma)
        const = 0;
15  };
```

Rather than make its own **PathsRecord_**, the SDE delegates that task to the **StepAccumulator_**; the latter can store any necessary information about the steps as they are made, before **NewPathsRecord** is invoked.

Sometimes the stepper has a need for additional random variables – *e.g.*, to handle a soft boundary in a Merton model, or to regularize the step in a Heston model – which cannot be predicted by **NumGaussians** for the Monte Carlo engine. The input **more_randoms**, created using **Random_::Branch**, is provided for this contingency. Any such use will not be repeatable in bumped scenarios, so it must be recorded in the paths record.

The Monte Carlo driver will ensure that the paths record endures longer than the workspace, so the workspace created by **NewWorkspace** is permitted to keep the input reference.

The **Envelope** method is used by the PDE to set up a grid; it gives upper and lower bounds to use for each state variable. **Envelope** is also called at the vol start time to find the initial state variables (though our convention is to use all zeroes).

The `StepAccumulator_` and `ModelStepper_` are constructed by the `SDE_` to its own specifications; they must communicate with each other, but each model can specialize that communication to its own ends. Similarly, if a `PathsRecord_` is used, it must be compatible with the accumulator (which constructs it) and with the steppers.

10.12   Valuation and Reevaluation

At many stages of the valuation process, particularly in Monte Carlo path generation, we must be aware of the difference between base and bumped valuation. Data of various kinds must be captured during the former, and reused in the latter, case. We reflect this by creating a *re-evaluator* to support repeated valuation:

```
──────────── Valuation.h ────────────
class ReEvaluator_ : noncopyable
{
protected:
    Vector_<pair<String_, double>> baseVals_;
public:
    virtual ~ReEvaluator_();
    virtual Vector_<pair<String_, double>> Values
        (_ENV, const Model_* bumped_model = nullptr)
    const = 0;
};
```

A call to `Values()` returns the base values. Evaluation using the base model is performed inside the constructor of a derived class will compute the values, which will also store any information necessary to compute stable bumped values.[12]

In some advanced distributed applications, there is a slight performance gain from making `ReEvaluator_` be `Storable_`, so that a base valuation on a single machine can be used to support bumped valuations on various remote machines. However, this is an onerous requirement which adds to the development effort for every new model. See Sec. 5.5 for the starting point of an example implementation.

---

[12]Storing the base values as a `const public` member would seem more natural, but makes the derived class implementation awkward because they are naturally the last, not the first, quantity available to the constructor.

10.13   Use Case Review: PDE

Let us see how these pieces fit together for a PDE pricing run. The flow of control proceeds as follows:

(1) The trade describes its underlyings.
(2) The model produces an SDE describing the dynamics of the necessary assets.
(3) The SDE produces a value request for use by the trade.
(4) The PDE solver wraps that value request in another, which will handle requests for past data and will provide stream locations to which values can be written.
(5) The trade produces a payout, filling out the value request in the process.
(6) The SDE produces an asset model which will translate state variables to the requested observable index values.
(7) The PDE solver stores the SDE's steppers between event times (possibly interpolating event times of its own to limit the step size), and computes the envelope it will use.
(8) The PDE solver translates the payout's backward induction steps to actions manipulating the node values.
(9) The inner loop is run for each time step; see below.
(10) The rollback terminates at the vol start time, regardless of the reset time; at this point each slab is collapsed to a single value.
(11) Backward induction actions before the vol start time, if any, are applied to the vector of values, and payments from event times before the vol start time, if any, are added.
(12) The solver rescales the per-stream values to account for the discounting from vol start time to value date.

The inner loop is a dance of information exchange between abstract objects. At each spatial node:

(1) The solver sends the state variables to the asset model.
(2) The asset model updates index values and returns an update token.
(3) The solver sends this token, and a node values accessor, to the payout.
(4) The payout uses index values to compute payment amounts, which it adds to the node values.
(5) The solver applies the backward induction actions.

If we have chosen to have the trade support PDE's more directly (see the discussion at the end of Sec. 10.5.3), then the payout can manipulate the node values directly rather than deferring all backward induction actions to the solver.

Note the use of the update token to ensure that actions are taken in the necessary order: an update token can only be obtained from an asset model, by giving it the state variables (coordinates in the PDE grid, transformed by a coordinate map) at the node being evaluated. The asset model ensures the market observables are ready for the payout to read, before returning the update token that allows the computation to continue.

Once the rollback is complete, we know the value of each stream, and the payout provides the information for translating these to trade values. PDE methods are generally stable, and the progress of a bumped run does not differ materially from that shown here; the main difference is that we will likely reuse the base envelope, especially if we are rescaling the grid during the run.

10.14   Use Case Review: Monte Carlo and Hedge

In a Monte Carlo, we must always distinguish between *base* and *bumped* valuations. The latter will reuse information from the base run in several ways, all aimed at preventing a discontinuous value change:

- To fix the order of eigenmodes, and other ordering issues.
- To fix AMC exercise decisions, which are otherwise chaotic.
- To repeat discrete jump or default events – we will reweight paths instead of generating different events.

The role of the paths record is to support the last of these needs, by making the history of the base valuation available during the bumped run. The Monte Carlo engine, which controls the AMC decider, is responsible for fixing decisions; this is accomplished by storing the base decisions in a fake decider which simply re-applies them. Eigenmode ordering and the like are handled by the SDE during creation of the stepper; for this purpose, the base stepper must be supplied so that a consistent bumped stepper can be created.

Given this, the flow of control proceeds as follows:

(1) The trade describes its underlyings.
(2) The model produces an SDE describing the dynamics of the necessary assets.
(3) The SDE produces a value request for use by the trade.

(4) The MC solver wraps that value request in another, which will handle requests for past data and will provide stream locations to which values can be written.

(5) The trade produces a payout, filling out the value request in the process.

(6) The SDE produces an asset model which will translate state variables to index values, and will populate index paths as necessary.

(7) The MC solver stores the model's steppers between event times; when creating steppers for the bumped run, it supplies the stored stepper from the base run.

(8) The trade is evolved forward to the vol start time, and the accumulator is used to find the starting state variables, once only (not per path).

(9) For event times in the future, the inner loop is run for each path; see below.

(10) Backward induction actions are applied to the whole path set. These change values by creating new nodes within their streams.

(11) The node values are discounted, averaged across paths, and summed to produce stream values.

(12) The payout provides the weights to translate to trade values.

The inner loop is changed from the PDE case exactly as one would expect. The `operator=` member of `Payout_::State_` lets us initialize the state appropriately at the start of each path. For each event time on the path:

(1) The solver invokes a stepper to change the model state.

(2) The solver sends the state variables to the asset model.

(3) The asset model updates index values and index paths, and returns an update token.

(4) The solver invokes `DoDefault` as necessary, then `DoNode`, using this token.

(5) The payout uses index values and paths to compute payment amounts, which it adds to the node values.

10.14.1 *Causality*

The inner loop displayed here implements a *causal* Monte Carlo, in which the thread of execution moves forward in simulated financial time so that information from the future can never be used in a payout. Based on the same protocols, we can also write an *acausal* or *whole-path* Monte Carlo, where we take all the model steps, then circle back and compute the resulting payout at each event time. The only design constraint is that the `ValueRequest_` must not reuse addresses across different event times.

Acausal Monte Carlo, while less aesthetically appealing, may be faster in practice because it keeps a given piece of code – the stepper or the payout – in-process longer, reducing cache-swapping costs.

10.15 Costs and Benefits

This is the reward for the rigorous abstraction process we have followed in this chapter: the two methods are generic to the largest feasible class of models and of trades, and both access the model and trade in the same way. This is our path toward enabling the most powerful models: anything which creates steppers and models assets is *ipso facto* a model, and can be used to price any trade.

Many pet projects are casualties of this approach. There is no place here for tightly coupled methods which compute one particular hedge (generally an equity or FX delta) directly within the Monte Carlo;[13] nor for highly model-specific methods such as PDEs for Asian options. Dependence on such highly specialized techniques is, in our opinion, a sign of a practitioner who has turned his back on the more important problem of computing a wide class of risks using a range of different models with maximum flexibility.

10.16 Assembling the Class Hierarchy

Here we work bottom-up, defining the most basic classes first. For a top-down view, where the motivation for a class is presented before its definition is known, it is advised to read this section backwards.

10.16.1 *Stepper*

Defined in Sec. 10.11.

---

[13]Sandeep Jain has successfully applied some adjoint methods within this general framework, computing efficient hedges for generic products in a reasonably broad class of models.

10.16.2   *Asset Values and Tokens*

The update token defined in Sec. 10.8 is a *view* of underlying data which
is manipulated by the model:

```
                            ——— AssetModel.h ———
struct PathFixings_ : noncopyable
{
    Vector_<> vals_;
    Vector_< shared_ptr<IndexPath_>> paths_;
    Vector_<Handle_<IndexPath_>> pathRO_;
    int valMask_, dateMask_;

    UpdateToken_ Token(const DateTime_& evt_t) const
    {
        return UpdateToken_(vals_.begin(), pathRO_.begin(),
            valMask_, dateMask_, evt_t);
    }
};
```

This object is just a container of data; it does not know the time, or monitor
how its contents are manipulated. It stores the index paths both as shared
pointers (to give write access to the asset model) and again as `Handle_s`
(to give read-only access to the payout).

10.16.3   *SDE*

The SDE produces steppers, which advance the state in a Monte Carlo or
supply coefficients to a PDE; and also provides the initial value request,
and the assets which communicate index fixings to the trade.

```
                            ——— SDE.h ———
class SDE_ : noncopyable
{
public:
    virtual ~SDE_();
    virtual ValueRequest_* NewRequest() const = 0;
    virtual Asset_* NewAsset(ValueRequest_& req) const = 0;

    virtual StepAccumulator_* NewAccumulator() const = 0;
    virtual ModelStepper_* NewStepper
        (const DateTime_& from,
         const DateTime_& to,
         StepAccumulator_* cumulative,
         ModelStepper_* exemplar)
        const = 0;
};
```

The `StepAccumulator_` input to `NewStepper` is that returned from `NewAccumulator`; it can be NULL if the model needs no cumulative information across times or paths. The `exemplar` will be non-NULL only in bumped valuations, when it will contain the base-case output from `NewStepper`.

10.16.4 *Model*

The model's role is to produce an SDE for pricing, once the trade underlying is known.

```
──────────────────────────── Model.h ────────────────────────────
class Model_ : public Storable_
{
    // single slide
    virtual Model_* Mutant_Model
5       (const String_* new_name = nullptr,
         const Slide_* slide = nullptr)
    const = 0;
public:
    virtual Handle_<SDE_> ForTrade
10       (_ENV, const Underlying_& trade)
    const = 0;

    virtual Handle_<YieldCurve_> YieldCurve
        (const Ccy_& ccy)
15   const = 0;

    virtual DateTime_ VolStart() const = 0;

    Model_* Mutant_Model
20       (const String_& new_name,
         const Vector_<Handle_<Slide_>>& slides)
    const;
};
```

We will discuss slides and `Mutant_Model` in Ch. 15. The access to the model's yield curves supports some closed-form pricing; see Ch. 14.

10.16.5 *Trade*

The trade must state its underlyings, and also (given a value request) produce the payout. It turns out that trades are not `Storable_`; we will discuss this in Ch. 11, next.

10.16.6   *Historical Data Access*

The `ValueRequest_` provided by the model should not have to concern itself with requests for historical data; and the trade, in making requests, should not even be aware of whether the resulting fixings are historical or simulated. Thus we need to intercede between the two, supplying historical fixings to trades.

────────────────────────── *ValueHistorical.h* ──────────────────────────

```
class PastAwareRequest_ : public ValueRequestImp_
{
    // supplied by the SDE_, to handle requests in the future:
    ValueRequestImp_& model_;
5
    const Environment_& env_;     // or other fixings access
    DateTime_ resetTime_;
    Date_ accountingDate_;

10  // implement the construction of a single historical path
    Handle_<IndexPathHistorical_> HistoricalPath
        (const Index_& index,
         const Handle_<IndexPathHistorical_>& prior);

15  struct History_ : IndexPathHistory_
    {
        PastAwareRequest_* parent_;
        const IndexPathHistory_* base_;
        History_(parent_, base_) {}
20
        DateTime_ ResetTime() const
        {
            return parent_->resetTime_;
        }
25      Handle_<IndexPathHistorical_> History
            (const Index_& index)
        const
        {
            auto prior = base_
30              ? base_->History(index)
                : Handle_<IndexPathHistorical_>();
            return parent_->HistoricalPath(index, prior);
        }
    };
35
    Handle_<Payment::Tag_> PayDst
        (const Payment_& flow)
    {
        return flow.date_ < accountingDate_
40          ? Payment::Null()
            : model_.PayDst(flow);
```

```
        }

        Handle_<Payment::Default::Tag_> DefaultDst
45          (const String_& stream)
        {
            return model_.DefaultDst(stream);
        }

50      address_t Fixing
            (const DateTime_& event,
             const Index_& index)
        {
            return event < resetTime_
55              ? model_.InsertFixing(index.Fixing(0, event))
                : model_.Fixing(event, index);
        }
        address_t InsertFixing(double fixing)
        {
60          return model_.InsertFixing(fixing);
        }


        IndexAddress_ IndexPath
            (const DateTime_& last_event_time,
65           const Index_& index)
        {
            History_ h(this, 0);
            return last_event_time < resetTime_
                ? model_.InsertPath
70                  (handle_cast<IndexPath_>(h.History(index)))
                : model_.IndexPath(last_event_time, index, &h);
        }
        IndexAddress_ IndexPath
            (const DateTime_& last_event,
75           const Index_& index,
             const IndexPathHistory_* historical)
        {
            History_ h(this, historical);
            return model_.IndexPath(last_event, index, &h);
80      }
        IndexAddress_ InsertPath
            (const Handle_<IndexPath_>& path)
        {
            return model_.InsertPath(path);
85      }
    };
```

It is not immediately obvious how `InsertFixing` and the three-argument
form of `InsertPath` can ever be called for this class. However, it is good

practice to have them simply forward to `model_`, so we will be able to nest `PastAwareRequest_` inside another decorator if need be.[14]

10.16.7 *Assets*

The asset is almost completely blank:

```
                        ____ Asset.h ____
class Asset_  : noncopyable
{
public:
    virtual UpdateToken_ Update
 5      (const DateTime_& event_time,
        const Vector_<>& state) = 0;
};
```

All our concrete assets will share an implementation scaffold, described in Sec. 13.3, which individual models will populate with model-specific local updaters.

10.16.8 *Solvers*

Each solver inherits from `ReEvaluator_`. The Monte Carlo solver exists in namespace `MonteCarlo`:

```
                        ____ MC.h ____
class Task_  : public ReEvaluator_
{
    scoped_ptr<ValueRequest_> request_;
    scoped_ptr<const Payout_> payout_;
 5  scoped_ptr<StepAccumulator_> cumulant_;
    Vector_<Handle_<ModelStepper_>> steps_;
    scoped_ptr<PathsRecord_> paths_;

public:
10  Task_(const SDE_& model,
            ValueRequest_* request,
            const Payout_* payout);
};
```

The `request` and `payout` are *orphan* pointers, whose memory belongs to the `MC_` once the constructor is called.[15]

We could pass the `Trade_` to the task, and let it internally use the trade and `SDE_` to make the value request and payout. However, that would

---

[14]For example, this is needed when forecasting expected payments, which is outside the scope of this volume.

[15]Passing bare pointers may be considered bad form; in practice we will immediately catch them in `scoped_ptrs`.

create a compilation dependency of Monte Carlo tasks on `Trade_`s, which we prefer to avoid; the two should be separate and equal.

This page intentionally left blank

Chapter 11

Standardized Trades

The fastest way to understand the protocols for numerical pricing is to look at a few trades, which of course will implement these protocols. We will return to pricing in those special cases where a closed form is available, in Ch. 14.

11.1 Trade Classes

We try to keep the trade's interface as narrow as possible. This preserves the maximum of implementation freedom, and more importantly, makes the task of adding a new trade type straightforward and relatively painless.[1]

```
──────────────────── Trade.h ────────────────────
class Trade_ : noncopyable
{
public:
    const Vector_<String> valueNames_;
    const Underlying_ underlying_;
    const Ccy_ valueCcy_;

    Trade_(valueNames_, underlying_, valueCcy_);

    virtual Payout_* MakePayout
        (const ValuationParameters_& parameters,
         ValueRequest_& value_request)
    const = 0;
};
```

---

[1] A repeated "anti-pattern" in financial software is a fat interface for trades, which incentivizes developers to extend existing trade types even when creating a new type would be more appropriate. This practice is self-reinforcing where it is used, so one or two trade classes become grossly bloated and no one dares to attempt breaking them up.

To avoid repeated evaluation of the underlying (and since there is no meaningful use of a trade that does not reference the underlying), it is stored in the base class upon construction. Note that there is no need for otiose accessor functions; `const public` data serves our purpose exactly and concisely. The `valueNames_` are the keys of `Payout_::StreamWeights` from Sec. 10.10; `Valuation::Parameters_` is a `settings` type – see Sec. 3.4 – containing discretization instructions which may be used by more complex trades.

The alert reader might wonder why a `Trade_` is not `Storable_`. We do not store the trade itself, but another object containing the information from which it was made:

```
────────────────── Trade.h ──────────────────
   class TradeData_ : public Storable_
   {
      mutable Handle_<Trade_> parsed_;
      virtual Trade_* XParse() const = 0;
 5 public:
      TradeData_(const String_& name) : Storable_("Trade", name) {}
      Handle_<Trade_> Parse() const;
      void Clear() const;    // un-Parse
      Bookkeeping_ Bookkeeping() const;
10 };
```

This is the "trade" seen at the public interface by library users. The `Bookkeeping_` contains non-financial information about the trade, such as the trade date, counterparty, or nominal notional. We will not treat these subjects in this volume.

`Parse` is called to convert the storable `TradeData_` to an (immutable) `Trade_`.[2] `Clear` may be called if necessary to free memory; parsed user-scripted trades are typically much larger in memory than their source scripts.

```
────────────────── Trade.cpp ──────────────────
   Handle_<Trade_> TradeData_::Parse() const
   {
      if (parsed_.Empty())
         parsed_.reset(XParse());
 5    return parsed_;
   }
```

The interface of `Trade_` is quite satisfactorily narrow: in fact, too narrow for many purposes. The missing functionality includes:

---

[2]To make this basic implementation thread-safe, we would need to add a `mutex` member datum in the trade data, and lock it here.

- More revealing representations than the rather opaque `Payout_`, for use in closed form pricing. We will address this with mixins; see Ch. 14.
- A query interface for introspection, *e.g.*, to identify range accrual trades.
- The `Storable_` interface itself, which would allow users to inspect the parsed `Trade_` object directly. The methods described in Sec. 5.5 can provide a partial fix.

11.2 Cash

The simplest possible derivative trade is an agreement to pay deterministic cash flows at some set of future dates. We need to implement this only for flows in a single currency; multi-currency flows can be formed as composite trades (see Sec. 11.9) or, more likely, booked as separate transactions.

Thus the trade's data during pricing are quite simple:

```
———————————————— CashTrade.cpp ————————————————
struct CashTrade_ : Trade_
{
    Vector_<Flow_> flows_;
```

We use `struct` rather than `class`, since we will keep the entire implementation inside a source file and publish only a factory function in the header. The other necessary data is stored directly in `Trade_`.

```
———————————————— CashTrade.cpp ————————————————
    CashTrade_(const String_& name, const Ccy_& ccy, flows_)
        :
        Trade_(V1(name), CashUnderlying(ccy, flows), ccy)
        { }
5
    Payout_* MakePayout
        (const ValuationParameters_&,
         ValueRequest_& value_request)
    const override;
10 };
```

The function `CashUnderlying` constructs an `Underlying_` with the appropriate currency and last payment date, extracted from the `flows`.

Now we can write the `Payout_` (see Sec. 10.10) which will be used in numerical pricing. (Using a numerical method for this purpose is obviously overkill; but ensuring its efficiency is a test of the quality of our framework.)

```
/*──────────────────── CashTrade.cpp ────────────────────*/
struct CashPayout_ : Payout_    // verbose implementation
{
    const String_ name_;    // of trade and of stream
    Vector_<pair<dst_t, double>> flows_;

    CashPayout_(const String_& name) : name_(name) {}

    Vector_<DateTime_> EventTimes() const
    { return V1(DateTime::Minimum()); }

    void DoNode
        (const UpdateToken_&, State_*, NodeValues_& pay_dst)
    const
    {
        for (auto f : flows_)
            pay_dst[f.first] += f.second;
    }

    weights_t StreamWeights() const
    {
        weights_t r;
        r[name_] = Vector::V1(make_pair(name_, 1.0));
        return r;
    }
};
```

The cashflows of the trade become **pairs** of tag and amount, since the tag
encapsulates all information except the amount (which is often stochastic)
about a trade's payment. The payments will be made into a stream whose
name is the same as the trade's; thus **StreamWeights** needs only to identify
that fact. We must create a nominal event time, which we set in the dim
past. We expect **DoNode** to be called for that event time (and only that
one), at which point we make all our payments.

The above class promises several restrictions – no backward induction,
a single event date, and a single stream for the trade – that are common
to most simple trades. So it is best to put them in an implementation
base class and reuse their code. We want to allow the potential to mix
and match these simplifications, rather than jump straight to the simplest case. In C++11, we can do this with forwarding constructors. For
instance, a simplifying wrapper to provide a single name for both trade
and stream has a template base class, and an variadic template forwarding
constructor:

```
_____ PayoutEuropean.h _____
template<class T_ = Payout_> class PayoutSingle_ : public T_
{
public:
    const String_& name_;
    Payout_::weights_t StreamWeights() const
    { return Payout::IdentityWeight(name_); }
protected:
    template<typename... Args_> PayoutSingle_(const String_& name,
        Args_&&... args) : T_(forward<Args_>(args)...),
        name_(name) {}
};
```

Similar wrappers define a payout with no backward induction, and the
subcase of a payout with only a single event time (thus no need for backward
induction).

```
_____ PayoutEuropean.h _____
template<class T_ = Payout_> class PayoutForward_ : public T_
{
protected:
    template<typename... Args_> PayoutForward_(Args_&&... args)
    : T_(forward<Args_>(args)...) {}
    Vector_<BackwardInduction::Action_> BackwardSteps() const
    { return Vector_<BackwardInduction::Action_>(); }
};

template<class T_ = Payout_> class PayoutEuropean_
    : public PayoutForward_<T_>
{
public:
    const DateTime_& eventTime_;
    Vector_<DateTime_> EventTimes() const
    { return Vector::V1(eventTime_); }
protected:
    template<typename... Args_> PayoutEuropean_
        (const DateTime_& event_time, Args_&&... args)
    : PayoutForward_<T_>(forward<Args_>(args)...),
        eventTime_(event_time) {}
};

using PayoutSimple_ = PayoutSingle_<PayoutEuropean_<>>;
```

Now `CashPayout_` can derive from `PayoutSimple_`, and needs to implement
only `DoNode`.

11.2.1 Setup of Payments

The value request is the tool that lets the trade create a payout, by providing it with promised addresses and payment tags. But we have to convert

the `Flow_` objects held by the trade into `Payment_` objects which communicate the context of the cashflow.

```
——————————— CashTrade.cpp ———————————
Payment_ MakePayment
   (const String_& stream,
    const Ccy_& ccy,
    const Flow_& flow)
{
    static const DateTime_ WHEN = DateTime::Minimum();
    static const Payment::Info_ INFO("Contractual cashflow", WHEN);
    return Payment_(WHEN, ccy, flow.payDate_, stream, INFO);
}
```

We could make this a member, rather than a nonmember to which class data are passed; but we habitually prefer to keep classes small and function interfaces explicit, even at some cost in verbosity.

```
——————————— CashTrade.cpp ———————————
Payout_* CashTrade_::MakePayout
   (const ValuationParameters_&,
    ValueRequest_& value_request)
const
{
    const String_& name = valueNames_[0];
    unique_ptr<CashPayout_> retval(new CashPayout_(name));
    for (auto f : flows_)
    {
        Payout_::dst_t tag = value_request.PayDst
                (MakePayment(name, valueCcy_, f));
        if (tag != Payment::Null())
            retval->flows_.push_back(make_pair(tag, f.amount_));
    }
    return retval.release();
}
```

This shows the use the value request, and of `Payment::Null` to eliminate worthless cashflows – the latter is an optimization[3] only, and not necessary for correct pricing or reporting.

```
——————————— CashTrade.cpp ———————————
/*IF-------------------------------------------------------
storable CashTradeData
    Deterministic cashflows only
version 1
&members
name is ?string
    Name of the trade
ccy is enum Ccy
```

---

[3]Not a very crucial one.

```
         Currency of cashflows
10  payDates is date[]
         Dates on which cashflows are paid
     amounts is number[]
         Cashflow amounts corresponding to dates
     &conditions
15  payDates.size() == amounts.size()
         Must have one cashflow amount for each payment date
     -IF----------------------------------------------------*/
```

This mark-up supports the trade data, which is extremely similar to the `CashTrade_` used for pricing:

```
──────────── CashTrade.cpp ────────────
    struct CashTradeData_ : TradeData_
    {
        Ccy_ ccy_;
        Vector_<Date_> dates_;
5       Vector_<> amounts_;

        Trade_* XParse() const override
        {
            return new CashTrade_
10              (name_, ccy_, Apply(ZipToFlow, dates_, amounts_));
        }

        void Write(Archive::Store_& dst) const
        {
15          CashTradeData_v1::XWrite(dst, name_, ccy_, dates_, amounts_);
        }

        CashTradeData_(const String_& name, ccy_, dates_, amounts_)
        :
20      TradeData_(name)
            {  }
    };
```

The duplication of members in `CashTrade_` and `CashTradeData_` is the price we pay for control over `Parse` and `Clear`.[4]

11.3   Equity and FX

A slightly more interesting trade is an equity forward – a contract to receive an equity index (or the cash value thereof) in exchange for a fixed amount

---

[4]If a large amount of data was shared, we might put it in a `Handle_` for use by both objects.

at a forward date. Now that we have seen the relationship of `Payout_`, `Trade_`, and `TradeData_`, we can write them from the inside out – this is how we will usually proceed when creating a new trade.

In this volume we do not make a distinction between trades and *instruments* (such as shares of stock), preferring to use the same concepts and classes for both. This betrays a background in over-the-counter derivatives.

11.3.1 *Equity Forward Payout*

What must the payout look like?

```
———————————————— EquityTrade.cpp ————
struct EquityForwardPayout_ : PayoutSimple_
{
    Valuation::address_t fixing_;
    double strike_, size_;    // size is signed
5   dst_t dst_;

    EquityForwardPayout_
      (const String_& name, const DateTime_& expiry,
       fixing_, strike_, size_, dst_)
10  : PayoutSimple_(name, expiry)
    {  }

    Vector_<DateTime_> EventTimes() const
    {
15      return Vector::V1(expiry_);
    }

    void DoNode
        (const UpdateToken_& values,
20       State_*,
         NodeValues_& pay)
    const
    {
        assert(values.eventTime_ == expiry_);
25      const double spot = values[fixing_];
        pay[dst_] += size_ * (spot - strike_);
    }
};
```

This illustrates the extraction of the necessary fixing – the index spot value at expiry – from the `values`.[5] The trade must now have the necessary data to construct this payout:

---

[5] Our generation of a single cash payment is equivalent to assuming cash settlement. A cash system might have to make a finer distinction, by treating the stock delivery the same as a cashflow.

```
————————————————— EquityTrade.cpp —————————————————
Payout_* EquityForward_::MakePayout
   (const ValuationParameters_&,
    ValueRequest_& mkt)
const override
{
   const String_& name = valueNames_[0];
   Handle_<Payment::Tag_> payDst(mkt.PayDst(MakePayment
      (expiry_, valueCcy_, name)));
   return new EquityForwardPayout_
         (name, expiry_, mkt.Fixing(expiry_, *index_),
          strike_, size_, payDst);
}
```

Here we have encapsulated two tasks – forming the index, and describing the payment – in the member `eqIndex_` and the function `MakePayment` respectively. The latter simply marshals other member data:

```
————————————————— EquityTrade.cpp —————————————————
Payment_ MakePayment
   (const DateTime_& expiry,
    const Ccy_& ccy,
    const String_& stream)
{
   Payment::Info_ info("Equity forward delivery", expiry);
   return Payment_(expiry, ccy, expiry.Date(), stream, info);
}
```

11.3.2 *Equity Index*

A spot equity price is a simple piece of market data, so of course we will have a class derived from `Index_` to represent it. It is best to allow for observation of equity forward as well as spot prices, though we do not need them for this trade.

```
————————————————— IndexEquity.h —————————————————
namespace Index
{
   class Equity_ : public Index_    // spot or forward equity price
   {
      Cell_ delivery_;    // empty, date, or increment
      String_ Name() const;
   public:
      const String_ eqName_;
      Date_ Delivery(const DateTime_& fixing_time) const;

      // can't supply both date and increment!
      Equity_(const String_& eq_name,
         const Date_* delivery_date = nullptr,
```

```
                   const String_* delay_increment = nullptr);
15      };
    }
```

As with the implied vol index in Sec. 9.4, we store either a fixed delivery date or a string representing an increment from event time to delivery, and call `Delivery` once we know the event time.

If we are using the index-naming conventions of Sec. 9.1, we must produce the expected name:

```
────────────── IndexEquity.cpp ──────────────
String_ Index::Equity_::Name() const
{
    String_ ret = "EQ[" + eqName_ + "]";
    if (Cell::IsString(delivery_))
5       ret += ">" + Cell::ToString(delivery_);
    else if (Cell::IsDate(delivery_))
        ret += "@" + Date::ToString(Cell::ToDate(delivery_));
    return ret;
}
```

The index parser should be capable of reconstructing the index from the resulting name.

11.3.3   *Equity Forward Data*

Since the index contains the equity name and payment date (once we supply the expiry), it is natural to supply it to the trade's constructor.

```
────────────── EquityTrade.cpp ──────────────
EquityForward_::EquityForward_
    (const String_& trade_name,
     const Handle_<Index::Equity_>& index,
     const Ccy_& ccy,
5    expiry_, strike_, size_)
     :
Trade_(V1(trade_name), EquityUnderlying(ccy, index, expiry), ccy),
index_(index)
    {    }
```

The `TradeData_` class whose `Parse` function produces an `EquityForward_` can store the index name, then parse it on demand to call the above constructor.

11.3.4 *FX Option*

The pricing of simple FX trades is conceptually similar to that of the corresponding equity trades, and the FX forward very closely resembles the equity forward just described. We have shown only cash settlement of equity forwards, but with FX we also have the option of physical settlement, making payments in two separate currencies rather than computing a value within the trade.

The payout code for these two possibilities, and also for the corresponding options, has enough common features that it is worth combining into a single class. Here the template parameter `T_` controls the payments actually made; it must provide an `operator()` taking the spot FX rate as input, and returning a pair of domestic and foreign payments to be made. Our `Payout_` calls a member of type `T_` within `DoNode`, then simply adds the resulting payments to the `NodeValues_`.

```
——————————— FxTrade.cpp ———————————
template<class T_> struct FxPayout_ : PayoutSimple_
{
    Valuation::address_t fixing_;
    dst_t domDst_, fgnDst_;
5   T_ payAmounts_; // spot -> (dom_pay_amt, fgn_pay_amt)

    FxPayout_
        (const String_& name, const DateTime_& expiry,
        fixings_, domDst_, fgnDst_, payAmounts_)
10  :
    PayoutSimple_(name, expiry)
    {   }

    void DoNode
15      (const UpdateToken_& values,
        State_*,
        NodeValues_& pay)
    const
    {
20      assert(values.eventTime_ == eventTime_);
        const double spot = values[fixing_];
        auto payDomFgn = payAmounts_(spot);
        if (!IsZero(payDomFgn.first))
            pay[domDst_] += payDomFgn.first;
25      if (!IsZero(payDomFgn.second))
            pay[fgnDst_] += payDomFgn.second;
    }
};
```

Now an FX forward trade is defined by the **execute_** function object, which also stores the trade's sign:

```
———————————————————— FxTrade.cpp ————————————
 template<bool CASH = false> struct ExecuteForward_
 {
     double domAmt_, fgnAmt_;    // signed

5    ExecuteForward_(double dom, double fgn, const RecPay_& rpf)
     : domAmt_(-dom * rpf.RecSign()), fgnAmt_(fgn * rpf.RecSign())
     { }

     pair<double, double> operator()(double spot) const
10   {
         return CASH
             ? make_pair(domAmt_ + fgnAmt_ * spot, 0.0)
             : make_pair(domAmt_, fgnAmt_);
     }
15 };
```

Physical settlement will generate a clearer payment report, and more accurately captures the nature of an FX forward. The receive/pay flag, by convention, refers to the foreign currency payment. Because **DoNode** will be called many times, we precompute signed payment amounts rather than store the flag.

A more complex **payAmounts_** calculator handles calls, puts and straddles; long or short options; and of course the underlying may still be a contract to pay or to receive the foreign currency. We precompute leverage flags for the in-the-money and out-of-the-money cases, and use the intrinsic value to decide which to use.

```
———————————————————— FxTrade.cpp ————————————
 template<bool CASH = false> struct ExecuteOption_
 {
     double domAmt_, fgnAmt_;
     double itmLev_, otmLev_;
5    ExecuteOption_(double dom_notional, double fgn_notional,
         const RecPay_& rp_fgn, const OptionType_& cps,
         int option_sign)
         :
     domAmt_(-dom_notional * rp_fgn.RecSign()),
10   fgnAmt_(fgn_notional * rp_fgn.RecSign()),
     itmLev_(option_sign * cps.Payout(1.0, 0.0)),
     otmLev_(-option_sign * cps.Payout(0.0, 1.0))
     { }

15   pair<double, double> operator()(double spot) const
     {
         const double intrinsic = domAmt_ + fgnAmt_ * spot;
         const double leverage = intrinsic > 0.0 ? itmLev_ : otmLev_;
```

```
20          return CASH
                ? make_pair(leverage * intrinsic, 0.0)
                : make_pair(leverage * domAmt_, leverage * fgnAmt_);
        }
};
```

At the public interface, however, we favor positive notional amounts and flags.

```
————————————————— FxTrade.cpp —————————————————
/*IF------------------------------------------------------------
storable FxOption
      Option to exchange currencies
version 1
5  &members
name is ?string
      Name of the trade
dom_ccy is enum Ccy
dom_amt is number
10      Domestic notional, positive
fgn_ccy is enum Ccy
fgn_amt is number
      Foreign notional, positive
rec_pay_fgn is enum RecPay
15      Whether foreign notional is received or paid upon exercise
expiry is datetime
      Time at which FX spot is observed and exercise decided
delivery is ?date
      Date on which FX is paid, if not deduced from expiry
20  &conditions
dom_amt > 0.0
fgn_amt > 0.0
-IF-----------------------------------------------------------*/
```

The **delivery** date is optional, because for standard options it can be deduced from the exercise date.

11.3.5 *Forcing Backward Induction*

We can write the payout another way, exercising the optionality during a backward induction sweep rather than immediately. This provides no implementation advantage, but illustrates the use of the backward induction protocols in a simple case.

```
————————————————— FxTrade.cpp —————————————————
struct FxOptionPayout_AMC_ : PayoutSingle_<>
{
    FxPayout_<ExecuteOption_<>> underlying_;
    int sign_;
```

```
 5      amount_t spotDst_;    // observable for AMC

       template<typename... Args_> FxOptionPayout_AMC_
         (const String_& name, sign_, spotDst_, underlying_...)
         :
10     PayoutSingle_(name)
       { }

       Vector_<DateTime_> EventTimes() const
       { return underlying_.EventTimes(); }
15
       void DoNode
           (const UpdateToken_& values,
           State_* state,
           NodeValues_& pay_dst)
20     const
       {
           underlying_.DoNode(values, state, pay_dst);
           pay_dst[spotDst_] = values[underlying_.fixing_];
       }
25
       Vector_<BackwardInduction::Action_> BackwardSteps() const
       {
           BackwardInduction::Action_ retval(underlying_.name_,
               underlying_.eventTime_, underlying_.eventTime_.Date());
30         BackwardInduction::Exercise_ bermEx;
           bermEx.sign_ = sign_;
           // allow exercise to observe the value stored by DoNode()
           bermEx.observables_.push_back(spotDst_);
           retval.details_ = bermEx;
35         return Vector::V1(retval);
       }
     };
```

The backward-induction action will terminate the stream when it gains value by doing so (assuming `sign_` is positive); its decision is based on a single observable, the spot FX price, which is all we need. There are no fees and no underlying streams received on exercise, because we are representing the trade as an FX forward with an option to terminate. The stream shares the trade's name and receives all the value, as expected by `PayoutSingle_`. In `MakePayment` when preparing to get the payment tags for the underlying (`domDst_` and `fgnDst_`), we will set the payment's `tag_.conditions_.exerciseCondition_` to `ON_CONTINUATION`, flagging that it is no longer an unconditional cashflow.

We could write the same trade as an option whose underlying is the cashflow in one currency (presumably the positive one), but where we must pay a fee denominated in the other currency to exercise. To accomplish this, we would

- Remove the payment from the value stream;
- Append the payment tag to `bermEx->fees_`;
- Set the payment condition to `ON_EXERCISE`.

Fee payments live in a kind of limbo, not assigned to any stream until they are inserted there by a Bermudan `Action_`.

11.4 Trade Amounts and Manipulators

To best support the creation of numerical payouts for swaps, we extend `PayoutForward_` to support the accumulation of leg-based flows. First we need a generic way to represent a (possibly) stochastic amount:

```
─────────────────── TradeAmount.h ───────────────────
class TradeAmount_ : noncopyable
{
public:
    virtual ~TradeAmount_();
    virtual double operator()
        (const UpdateToken_& values) const = 0;
};
```

The most common amount is simply a wrapper around a single fixing. We also provide a helper function to create this wrapper from a promise made by the asset request.

```
─────────────────── TradeAmount.h ───────────────────
namespace TradeAmount
{
    struct Fixing_ : TradeAmount_
    {
        const Valuation::address_t loc_;
        Fixing_(loc_) {}
        double operator()(const UpdateToken_& values) const
        {
            return values[loc_];
        }
    };
    inline Handle_<TradeAmount_> AsAmount
        (const Valuation::address_t& loc)
    { return new Fixing_(loc); }
```

More complex amounts will be needed, *e.g.*, for payments incorporating a margin or spread. We create a representation for a sum or product, including a deterministic part:[6]

```
                       ——— TradeAmountUtils.h ———
    template<class OP_> struct Combined_ : TradeAmount_
    {
        Vector_<Handle_<TradeAmount_>> stochastic_;
        double deterministic_;
5       Combined_(stochastic_, deterministic_) {}
        double operator()(const UpdateToken_& values) const
        {
            double retval = deterministic_;
            for (auto s : stochastic_)
10              retval = OP_()(retval, (*s)(values));
            return retval;
        }
    };
```

A `Combined_` amount efficiently computes a sum or product, but we must
marshal the arguments before calling its constructor. We provide an accumulator for this purpose: it holds and updates the inputs to the `Combined_`,
allowing us to accrete either stochastic or deterministic quantities. At the
end of the accumulation process, `NewAmount` constructs the `TradeAmount_`
we need.

```
                       ——— TradeAmountUtils.h ———
    template<class OP_> struct Accumulate_
    {
        Vector_<Handle_<TradeAmount_>> stochastic_;
        double deterministic_;
5
        Accumulate_(const Handle_<TradeAmount_>& s, deterministic_)
        {
            if (s)
                stochastic_.push_back(s);
10      }

        Accumulate_<OP_> operator()(const Handle_<TradeAmount_>& s)
        {
            Accumulate_<OP_> retval(*this);
15          retval.stochastic_.push_back(s);
            return retval;
        }
        Accumulate_<OP_> operator()(double d)
        {
20          Accumulate_<OP_> retval(*this);
            retval.deterministic_ = OP_()(retval.deterministic_, d);
            return retval;
        }
```

---

[6]We also have a degenerate `TradeAmount_` subclass for deterministic quantities, but it
is more efficient to avoid wrapping known values in the first place.

```
25          TradeAmount_* NewAmount() const
            { return new Combined_<OP_>(stochastic_, deterministic_); }
         };

         inline Accumulate_<plus<double>> Sum
30           (const Handle_<TradeAmount_>& s = Handle_<TradeAmount_>())
         {
             return Accumulate_<plus<double>>(s, 0.0);
         }
         inline Accumulate_<multiplies<double>> Product
35           (const Handle_<TradeAmount_>& s = Handle_<TradeAmount_>())
         {
             return Accumulate_<multiplies<double>>(s, 1.0);
         }
     }
```

The helper functions Sum and Product provide appropriately initialized accumulators for these purposes; Sec. 11.7 shows a use case. We will also want helper functions such as Difference and Ratio, but they must take precisely two arguments.

With this little library of manipulators, we can easily construct the computed quantities, such as accrued interest over a period, which are found in swaps.

11.5 Legs and Swaps

We begin with the description of a fixed leg trade, alluded to in Sec. 8.5.2. The trade data is:

LegBased.h

```
/*IF----------------------------------------------------------
storable FixedLegTrade
version 1
&members
5 name is ?string
ccy is enum Ccy
terms is handle LegScheduleParams
     Parameters from which leg is built
recPay is enum RecPay
10    Sign of flows
notional is number[]
     Single or per-period notional
couponRate is number[]
     Single or per-period coupon
15 notionalExchange is ?enum NotionalExchange
     States whether/when notionals are exchanged
-IF----------------------------------------------------------*/
```

As is our practice, we will design this class bottom-up, beginning with the components of its `Payout_`. We define a coupon payment, then a `Payout_` composed of coupons indexed by their event times:

```
                      ——— LegBased.h ———
namespace LegBased
{
    struct Coupon_
    {
        DateTime_ eventTime_;
        Handle_<TradeAmount_> rate_;
        double dcf_;
        Handle_<Payment::Tag_> pay_;
    };

    class Payout_ : public PayoutForward_<PayoutSingle_<>>
    {
        multimap<DateTime_, Coupon_> coupons_;
    public:
        Payout_(const String_& trade_name)
            : PayoutForward_<PayoutSingle_<>>(trade_name) {}

        Payout_& operator+=(const Coupon_& c);

        Vector_<DateTime_> EventTimes() const
        {
            return Unique(Keys(coupons_));
        }

        void DoNode
            (const UpdateToken_& vls,
             State_* state,
             NodeValues_&dst)
        const
        {
            auto tt = coupons_.equal_range(vls.eventTime_);
            for (auto pc = tt.first; pc != tt.second; ++pc)
            {
                const Coupon_& c = pc->second;
                dst[c.pay_] += c.dcf_ * (*c.rate_)(vls);
            }
        }
    };
```

Now forming a numerical payout, for any leg-based trade, is just a matter of creating the `Coupon_` amounts. We simplify this task with a support function which converts a `LegPeriod_` (Sec. 8.5) to a `LegBased::Coupon_`.

In order that this function be extensible to unforeseen `CouponRate_` types, we spread its implementation over two classes (in `namespace LegBased`):

```
——————————— LegBased.h ———————————
    struct MakeRate_ : noncopyable
    {
       virtual pair<DateTime_, Handle_<TradeAmount_>> operator()
          (ValueRequest_& request,
 5            const CouponRate_& rate)
       const;
    };

    struct MakeCoupon_
10   {
       ValueRequest_& valueRequest_;
       const Handle_<MakeRate_> makeRate_;
       const String_ tradeName_;
       const Ccy_ payCcy_;
15
       virtual ~MakeCoupon_();
       MakeCoupon_
          (ValueRequest_& v, makeRate_, tradeName_, payCcy_);

20      virtual Coupon_ operator()
          (const LegPeriod_& period)
       const;
    };
}
```

The default implementation `MakeRate_::operator()` recognizes the rate types – fixed, Libor, Libor plus margin, and interpolated stub Libor – discussed in Secs. 8.5 and 8.5.1. Other trades can reuse `MakeCoupon_` by defining their own classes, derived from `MakeRate_`, which will handle the trade's own rate type. Most rate types can be handled this way, without the need to override the implementation of `MakeCoupon_::operator()`.

So that basic rate functionality can be accessed without explicitly creating a `MakeRate_` object – which is stateless anyway – we provide the utility unary `+` operator:

```
——————————— LegBased.cpp ———————————
const LegBased::MakeRate_& operator+
   (const Handle_<LegBased::MakeRate_>& mr)
{
   static const LegBased::MakeRate_ DEFVAL;
 5 return mr.Empty() ? DEFVAL : *mr;
}
```

Thus we can call **+makeRate_** to get the default implementation if no override is specified. This simplifies the implementation of **MakeCoupon_** as well as of derived **MakeRate_** classes.

```
──────────────── LegBased.cpp ────────────────
   LegBased::Coupon_ LegBased::MakeCoupon_::operator()
       (const LegPeriod_& period)
   const
   {
5      Coupon_ ret;
       ret.dcf_ = period.accrual_->dcf_;
       Payment_ pay;
       pay.tag_.period_ = *period.accrual_;
       pay.tag_.description_ = "Coupon payment";
10     pay.ccy_ = payCcy_;
       pay.date_ = period.payDate_;
       pay.stream_ = tradeName_;
       tie(pay.eventTime_, ret.rate_) =
           (+makeRate_)(valueRequest_, *period.rate_);
15     ret.eventTime_ = pay.tag_.knownTime_ = pay.eventTime_;
       ret.pay_ = valueRequest_.PayDst(pay);
       return ret;
   }
```

11.5.1 *Putting it Together*

With this definition of a leg, a fixed or Libor leg trade contains just a leg, plus a value currency:

```
──────────────── LegTrade.cpp ────────────────
   class FixedLeg_ : public Trade_
   {
       Vector_<LegPeriod_> leg_;
       Ccy_ ccy_;
5  public:
       Payout_* MakePayout
           (const ValuationParameters_&,
            ValueRequest_& value_request)
       const override
10     {
           static const Handle_<LegBased::MakeRate_> DUMMY;
           unique_ptr<LegBased::Payout_> retval
               (new LegBased::Payout_(valueNames_[0]));
           LegBased::MakeCoupon_ makeCoupon
15             (value_request, DUMMY, valueNames_[0], ccy_);
           for (auto p : leg_)
               *retval += makeCoupon(p);
           return retval.release();
       }
20 };
```

We supply an empty `MakeRate_` object, content to let it be overridden by `MakeCoupon_`'s call to `operator+`. The coupons, once made, are accumulated by `operator+=` in the `Payout_`; once the latter is fully populated, it will compute the realized payment amounts at each node and supply them to the `NodeValues_`.

As usual, during the process of populating the `Payout_`, we gather enough information for the model and numerical solver to know their responsibilities as well; this communication task is conveniently encapsulated in `MakeCoupon_`.

Precisely the same approach works for Libor legs, with or without stubs and margins. The trade's only task is to mediate between the leg-construction functions of Sec. 8.5 and the `LegBased` payout.

11.6 Caps

To construct a cap, we first need to introduce the rate it pays:[7]

```
―――――――――――――――――――――― Cap.h ――――――――――――――――――――――
struct CapRate_ : CouponRate_
{
    Handle_<CouponRate_> underlying_;
    double strike_;
5   CapFloor_ type_;
    BuySell_ sign_;
};
```

Now we provide a derived class of `MakeRate_` which will handle this. We would like to handle caps on anything, not just on Libor, which complicates the implementation slightly.

```
―――――――――――――――――――――― Cap.h ――――――――――――――――――――――
struct MakeRateCapped_ : LegBased::MakeRate_
{
    Handle_<MakeRate_> base_;
    MakeRateCapped_(base_ = Handle_<MakeRate_>()) {}
5
    pair<DateTime_, Handle_<TradeAmount_>> operator()
        (ValueRequest_& model, const CouponRate_& underlying_rate)
    const override;
};
```

---

[7]The type `CapFloor_` is another machine-generated enumeration. We must be careful not to confuse caps with calls generally; a cap is a call on rates but a put on bonds.

```
———————————————— Cap.cpp ————————————————
pair<DateTime_, Handle_<TradeAmount_>> MakeRateCapped_::operator()
   (ValueRequest_& model,
     const CouponRate_& rate)
const
{
    if (auto cap = dynamic_cast<const CapRate_*>(&rate))
    {
        auto temp = (+base_)(model, *cap->underlying_);
        Handle_<TradeAmount_> capAmt
                (NewCapAmount
                        (temp.second, cap->strike_, cap->type_,
                        cap->sign_));
        return make_pair(temp.first, capAmt);
    }
    return (+base_)(model, rate);
}
```

A default-constructed `MakeRateCapped_` will support caps on Libor, again using `operator+` to supply the default implementation. To cap a more exotic rate, we need only provide `MakeRateCapped_`'s constructor with the engine (derived from `MakeRate_`) which computes fixings for that rate.

The construction of a Libor cap trade should now be entirely predictable. We begin with the trade data:

```
———————————————— Cap.cpp ————————————————
/*IF-----------------------------------------------------------
storable LiborCapTrade
    Cap/floor on Libor
version 1
&members
name is ?string
ccy is enum Ccy
terms is handle LegScheduleParams
    From which leg is generated
fixingHols is string
    If omitted, currency defaults are used
fixingDelay is ?integer[]
    A single constant, or a delay for each fixing
buySell is enum BuySell
capFloor is enum CapFloor
notional is number[]
    A single constant, or a notional for each pay period
strike is number[]
    A single constant, or a strike for each period
-IF----------------------------------------------------------*/
```

This is sufficient data to call a function `LegBuild::LiborCap`, which will in turn call `LegBuild::Libor` (see Sec. 8.5) and then build a new leg with `CapRate_`s replacing the Libor rates.

11.7 Swaps and Swaptions

There is not much more to say about swaps: they are collections of one or more (but almost always two) legs, for each of which we will store data, build a leg, and add its payments to `LegBased::Payout_` for pricing. We may choose to implement swaps as a subclass of composite trade (see Sec. 11.9); this would give more visible separation of the contributions of each leg to the value, at the cost of increasing the number of streams in each pricing task.

A first cut of a European swaption would build the underlying swap as another `LegBased::Payout_`, then implement the option on this underlying during a backward induction phase (in Monte Carlo) or postprocessing step (in PDE); this would resemble the FX option payout in Sec. 11.3.5. However, this involves simulating the model state at each Libor fixing date of the underlying swap; thus it will be far faster, and also more precise, to create a payout which forecasts the Libor rates at the swaption's expiry.[8]

If the underlying swap is known to be standard, so that its rate and sensitivity can be described as indices (as in Sec. 9.1), then we can assign their computation to the model and write a very simple payout. This also allows optimizations on the model side (see Sec. 13.3.5 for an example), and is necessary for cash-settled swaptions (for which we do not show an implementation in this volume).

In general, though, we must consider options on swaps with time-varying coupons or notionals, which cannot be passed off to the model. We need to use the methods of Sec. 11.5, but creating *forecasts* of Libor at the swaption expiry rather than waiting for *fixings*. For brevity, in this section we ignore notionals, which add no truly new features to the problem.

Fortunately, we can accomplish the necessary forecasing with an overload of `LegBased::MakeRate_`:

```
―――――――――――――― Swaption.cpp ――――――――――――
struct MakeForecast_ : LegBased::MakeRate_
{
    DateTime_ expiry_;
```

---

[8]This is less profound than it sounds. We simply wish to value the underlying analytically at expiry, rather than by backward induction, if possible.

```
     MakeForecast_(expiry_) {}
 5   pair<DateTime_, Handle_<TradeAmount_>> operator()
       (ValueRequest_& req, const CouponRate_& rate) const
     {
         if (auto libor = dynamic_cast<const LiborRate_*>(&rate))
         {
10           Index::IR_ index(libor->ccy_, libor->tenor_);
             return make_pair(expiry_,
                 TradeAmount::AsAmount(req.Fixing(expiry_, index)));
         }
         return LegBased::MakeRate_::operator()(req, rate);
15   }
};
```

Now calling **LegBased::MakeCoupon_** with a **MakeForecast_** will convert the legs of the underlying swap into a series of **TradeAmount_** handles. This is still not good enough for our purposes: rather than pay the various coupons into a stream, we must obtain and sum their discounted values. Discounting on the trade side is somewhat unusual, but does allow more efficient support of some kinds of optionality.

Thus we replace **MakeCoupon_** altogether, with a function which will create discounted values for us to accumulate:

Swaption.cpp

```
struct ToDiscountedValue_
{
    ValueRequest_& valueRequest_;
    const String_ tradeName_;
 5  const Ccy_ payCcy_;
    const MakeForecast_ makeRate_;

    TradeAmount_* operator()(const LegPeriod_& pd) const
    {
10      Index::DF_ df(payCcy_, pd.payDate_);
        Valuation::address_t dfLoc =
            valueRequest_.Fixing(makeRate_.expiry_, df);
        auto product = TradeAmount::Product
                (makeRate_(valueRequest_, *pd.rate_).second)
15              (pd.accrual_->dcf_)
                (TradeAmount::AsAmount(dfLoc));
        return product.NewAmount();
    }
};
```

This functor creates **TradeAmount_s** for the rate, accrual fraction, and discount factor, then multiplies them together (using **Product** from

Sec. 11.4) to return a final **TradeAmount_** which can be evaluated at each node.[9]

A numerical swaption trade contains a list of such values, which represent the underlying:

```
——————————— Swaption.cpp ———————————
struct SwaptionPayout_ : PayoutSimple_
{
    int sign_;    // buy/sell
    Vector_<Handle_<TradeAmount_>> underlyings_;
    dst_t payDst_;

    SwaptionPayout_(const String_& name, const DateTime_& expiry,
        sign_, underlyings_, payDst_)
        :
    PayoutSimple_(name, expiry)
    { }

    void DoNode
        (const UpdateToken_& values,
        State_*,
        NodeValues_& pay)
    const override
    {
        assert(values.eventTime_ == eventTime_);
        double pv = 0.0;
        for (auto u : underlyings_)
            pv += (*u)(values);
        pay[payDst_] += sign_ > 0.0 ? Max(0.0, pv) : Min(0.0, pv);
    }
};
```

11.8 Bermudans

Pricing European swaptions in this way almost prepares us for Bermudan options. The only remaining step is to set up the backward induction actions which will support American Monte Carlo pricing. We must decide what observables to supply – including an indicator of the state of volatility, if we are using a stochastic-vol model – and form the backward induction actions as described in Sec. 10.9. Fortunately, our American Monte Carlo engine is designed to work robustly with fairly few observables (generally two to four); see Sec. 7.10.

---

[9]To be maximally pedantic, **Product** does not multiply, but returns a **TradeAmount** which will.

11.8.1 *Two Views*

This above procedure is natural if we view a Bermudan option as a union of European swaptions. Another, equally valid approach is to price a callable swap directly: we use the methods of Sec. 11.5 to define the swap payments, and then insert backward induction actions which, upon exercise, terminate the payment stream.

There is no strong reason to prefer either of these approaches over another. The union-of-swaptions method, since it forecasts exercise values at each exercise date, reduces numerical noise at the cost of increased computation per node. We should be aware of both approaches, so that we can price more complex callable swaps directly without the need to artificially divide them into swap and option components.

11.9 Composites

So far we have considered trades as independent entities, each producing a single value. There are several reasons we might need a more general view:

- A trade which makes identical payments, but on varying notional amounts, to many counterparties (*e.g.*, retail investors);
- A sum of trades, to be considered as a single trade for some purposes (*e.g.*, counterparty analysis);
- A collection of trades, to be simultaneously valued in a single Monte Carlo run for optimization.

We implement all such composites using two classes: a *collection trade* which gathers many trades into a single entity, and a *remapping trade* which changes the assignment of stream values to trade values. Two of our example use cases correspond directly to these classes; the other, a trade formed from a sum of component trades, is accomplished by first collecting and then remapping to combine the values.

11.9.1 *Remapping Trades*

To remap the components of a trade's value, we must preserve all the base trade's functionality but change the `StreamWeights`.

We provide a general linear mapping from source to destination values, with an arbitrary matrix of coefficients.

```
                    ———————— TradeComposite.cpp ————————
   /*IF------------------------------------------------------
   storable RemappingTrade
        Sums, and optionally redirects, values of other trades
   version 1
 5 &members
   name is ?string
        Name of the trade object
   base is handle TradeData
        A trade (possibly composite) providing the input values
10 src_names is ?string[]
        If supplied, names of the source values which are to be used;
        otherwise all value_names of base trade are used
   dst_names is ?string[]
        If supplied, names of the values produced by this trade;
15      otherwise trade name is used
   value_scalers is number[][]
        Coefficients in the linear combination,
        addressed as value_scaler[i_src][i_dst]
   &conditions
20 value_scalers.Rows() == \
   (src_names.empty() ? base->valueNames_ : src_names).size()
        Rows in linear mapping must correspond to src_names
   value_scalers.Cols() == Max(1, dst_names.size())
        Columns in linear mapping must correspond to dst_names
25 -IF------------------------------------------------------*/
```

What little implementation this requires is quite straightforward. The Payout_ forwards all requests to the base:

```
                    ———————— TradeComposite.cpp ————————
   struct LinearCombinationPayout_ : PayoutDecorated_
   {
        map<String_, map<String_, double>> coeffs_;
        LinearCombinationPayout_
 5          (const Handle_<Payout_>& base, coeffs_)
            :
        PayoutDecorated_(base)
        {   }

10      weights_t StreamWeights() const
        {
            weights_t retval, base = base_->StreamWeights();
            for (auto srcCoeff : coeffs_)
            {
15           NOTICE(srcCoeff.first);
             auto src = base.find(srcCoeff.first);
             REQUIRE(src != base.end(), "Source name not found in
                    trades underlying linear combination");
             for (auto dstCoeff : srcCoeff.second)
```

```
20          for (auto srcTerm : src->second)
               retval[dstCoeff.first].push_back(make_pair
                   (srcTerm.first,
                    dstCoeff.second * srcTerm.second));
            return retval;
25       }
         return retval;
      }
   };
```

Note the use of `PayoutDecorated_`, the null decorator of `Payout_`, as the base class; this supplies all the forwarding functions we would otherwise have to write repeatedly, as described in Sec. 2.6.

The `TradeData_` object whose contents are given by the mark-up just forwards them to a `RemapTrade_` implementation object, whose only real job is to translate the coefficients to the form required by the payout. The `base_` is usually itself a collection of component trades.

TradeComposite.cpp

```
struct TradeRemap_ : Trade_
{
   Handle_<Trade_> base_;
   Vector_<String_> srcNames_;
5  map<String_, map<String_, double>> remap_;

   TradeRemap_
       (const String_& name, base_,
       const Vector_<String_>& src_names,
10      const Vector_<String_>& dst_names,
       const Matrix_<>& coeffs_)
         :
       Trade_(dst_names.empty() ? Vector::V1(name) : dst_names,
           base_->underlying_, base_->valueCcy_),
15    srcNames_(src_names.empty() ? base_->valueNames_ : src_names)
       {
         for (int ii = 0; ii < srcNames_.size(); ++ii)
            for (int jj = 0; jj < valueNames_.size(); ++jj)
               if (!IsZero(coeffs_(ii, jj)))
20                remap_[srcNames_[ii]][valueNames_[jj]] =
                       coeffs_(ii, jj);
       }

   Payout_* MakePayout
25       (const ValuationParameters_& params,
       ValueRequest_& model)
       const override
       {
         Handle_<Payout_> base(base_->MakePayout(params, model));
30       return new LinearCombinationPayout_(base, remap_);
       }
   };
```

In addition to this very generic interface, we will provide special-purpose factory functions for common tasks. For instance, the simple sum trade mentioned at the start of this section is now quite easy:

```
────────────────────── TradeComposite.cpp ──────────────────────
   TradeData_* Trade::NewSum
      (const String_& name,
       const Vector_<Handle_<TradeData_>>& trades)
   {
5      if (trades.size() > 1)
       {
          // collect into a composite and sum that
          Handle_<TradeData_> temp
                 (new CollectionTradeData_(name, trades));
10         return NewSum(name, Vector::V1(temp));
       }
       auto base = trades[0];
       Matrix_<> coeffs(base->Parse()->valueNames_.size(), 1);
       coeffs.Fill(1.0);
15     return new RemappingTrade_(name, base,
              Vector_<String_>(), Vector_<String_>(), coeffs);
   }
```

11.9.2   *Collections*

True composites, containing several trades, require a little more implementation effort. The data model is extremely simple:

```
────────────────────── TradeComposite.cpp ──────────────────────
   /*IF----------------------------------------------------------
   storable CollectionTrade
        Collects values of other trades
   version 1
5  &members
   name is ?string
        Object's name
   trades is +handle TradeData
        All the trades which need to be valued for the composite
10 -IF--------------------------------------------------------*/
```

The **TradeData_** and **Trade_** are now very easy to implement, but the **Payout_** has some subtleties.

First, what if two component trades make payments into the same stream – or, more accurately, if the names of two streams coincide?[10] We cannot let the streams mix, which would confuse the trade values and most likely lead to double-counting. Thus during creation of the member payouts, we intervene in the requests made by member trades: for each, we add a unique prefix to the stream name before it is seen by the value request.

```
──────────────────── TradeComposite.cpp ────────────────────
struct XValuesForComposite_ : ValueRequest_
{
    ValueRequest_& base_;
    String_ prefix_;
5
    XValuesForComposite_(ValueRequest_& vr) : base_(vr) {}

    Handle_<Payment::Tag_> PayDst
        (const Payment_& flow)
10  {
        Payment_ temp(flow);
        temp.stream_ = prefix_ + temp.stream_;
        return base_.PayDst(temp);
    }
15
    Handle_<Payment::Default::Tag_> DefaultDst
        (const String_& stream)
    {
        return base_.DefaultDst(prefix_ + stream);
20  }
    // ...
};
```

The stream prefixes are decided by the `CollectionTrade_` during creation of its payout, so we can also store them in the payout. We will likely use a simple system of prefixes such as `"[0]"`, `"[1]"`, etc.[11]

The collection's state must store substates for every member payout. We simplify this with a template `Composite_` class, which derives from its template parameter[12] and contains a vector of shared pointers to elements of that base type.

---

[10]We have so far used only streams with the same name as the trade, but this is not a universal law. In scripted trades, for instance, our users will choose their own stream names.

[11]In some advanced applications we might have to wrap a composite around a preexisting payout, which itself may or may not be composite; to support this, we must put more unique information into the prefixes so that we can safely handle their nesting or absence.

[12]More precisely, from the non-`const` part of the parameter.

```
──────────── CompositeTrade.cpp ────────────
struct CollectionPayout_ : Composite_<const Payout_>
{
    Vector_<Vector_<DateTime_>> eventTimes_;
    Vector_<String_> streamPrefixes_;

    // handle state with a Composite_
    typedef Composite_<State_> state_t;
    State_* NewState() const override
    {
        unique_ptr<state_t> retval(new state_t);
        for (auto t : contents_)
            retval->Append(t->NewState());
        return retval.release();
    };

    void StartPath(State_* _state) const override
    {
        state_t& state = CoerceComposite(_state);
        assert(state.Size() == contents_.size());
        for (int it = 0; it < contents_.size(); ++it)
            contents_[it]->StartPath(state[it]);
    };
```

A second issue is the protection of the member payouts' event times; we must ensure that their DoNode functions are called only at the appropriate times. Thus we implement DoNode with preliminary checking:

```
──────────── CompositeTrade.cpp ────────────
    void DoNode
        (const UpdateToken_& vls,
         State_* _state,
         NodeValues_& pay_to)
    const override
    {
        state_t& myS = CoerceComposite(_state);
        assert(myS.Size() == contents_.size());
        const DateTime_& t = vls.eventTime_;
        for (int it = 0; it < contents_.size(); ++it)
        {
            if (BinarySearch(eventTimes_[it], t))
                contents_[it]->DoNode(vls, myS[it], pay_to);
        }
    }
```

An alternative, more efficient implementation avoids the BinarySearch by storing in the State_ an iterator into each component's eventTimes_. These iterators can be reset in StartPath, but care is needed: they cannot be reset to the respective **begin** iterators, since the component payouts may have events before the global reset time.

DoDefault works similarly, but instead of checking the presence of the event time we check that the default date is before the component's last event time.

When defining the stream weights, we must reuse the prefixes which our custom **ValueRequest_** added to the component stream names.

```
                              CompositeTrade.cpp
    weights_t StreamWeights() const
    {
        weights_t retval;
        for (int it = 0; it < contents_.size(); ++it)
        {
            for (auto sub : contents_[it]->StreamWeights())
            {
                REQUIRE(!retval.count(sub.first),
                  "Duplicate trade name in collection");
                auto prepend = [&](const pair<String_, double>& p)
                  ->pair<String_, double> { return make_pair
                  (streamPrefixes_[it] + p.first, p.second); };
                retval[sub.first] = Apply(prepend, sub.second);
            }
        }
        return retval;
    }
};
```

The marshalling of backward induction actions proceeds similarly; the prefix for each trade must be prepended to that trade's internal stream names, to create stream names which are seen by the numerical pricer.

Tools for semianalytic pricing of these trades will be discussed in Sec. 14.5.

Chapter 12

Curves

The word *curve* is a term of art, referring to a deterministic function of time used in pricing some set of trades. The prototypical example is the *yield curve*, originally referring to bond yields as a function of maturity but now used for forward Libor rates.

One subtle difference between these usages is worth noting. The bond yields are essentially "raw" market data, depending only on the quoted price in an unambiguous way. Forward rates fitted to some set of quoted instruments, however, are not directly quoted and not fully defined (also see Sec. 7.4). As a rule, we will focus on this latter type of curve.

12.1 Risk

When computing risk, there are two quantities we might demand: the sensitivity to the curve points (forwards) themselves, or to the market quotes used to build the curve. Clearly the relationship between the two is a function of the curve build method, since that method defines how the forwards will respond to a changed quote.

There are three ways to define a curve risk:

- *Store* the curve build method as part of the curve, then apply it to generate bumped curves (for the initial build instruments, or some other specified instruments).
- *Specify* a rebuild method as a required part of the risk specification, thus decoupling the initial build from the risk.
- *Ignore* the build method; construct tractable bumps, such as a constant change to the forwards within a given interval, and compute instrument exposures by inverting the Jacobian of quoted rates to these bumps.

The "store" method, or the degenerate case where only one build method is available, is most commonly used for yield curves; the "ignore" method (also called "bucket hedge") seems most common for ATM volatility curves.

The latter method has the disadvantage that p/l explanation cannot be as precise, because the details of the response to input quotes are not captured. However, the response functions for many common curve build algorithms are of low quality, and it may not be desirable to hew too closely to them. For instance, in yield curve building by linear interpolation of zero-coupon yields from spot, the later rates within an interval between instrument maturities respond more to a bump than do the earlier ones; this will inevitably be reflected in the risk when this method is used to rebuild bumped curves.

In what follows, we choose to ignore the build method; we regard this as superior to storing it. The additional code needed to specify a rebuild method is beyond this volume's scope.

12.2   Yield Curves

Libor forward curves, and the discount rates that underpin all derivatives pricing, are generally built as a unitary "yield curve." If we consider discounting at Libor, or at a fixed spread to Libor, then we need to build only a single curve – *i.e.*, our yield curve will have only a single functional degree of freedom. But this is a feature of the build method, not of the result; thus we will be careful to avoid making such a promise in our yield curve interface.

The main tasks of a curve are *forecasting* and *discounting*. We separate these in the interface:

```
                              YC.h
class YieldCurve_ : public Storable_
{
public:
    const Ccy_ ccy_;
5   YieldCurve_(const String_& name, const String_& ccy);
    virtual const DiscountCurve_& Discount
        (const CollateralType_& collateral)
    const = 0;
    virtual double FwdLibor
10      (const PeriodLength_& tenor, const Date_& fixing_date)
    const = 0;
};
```

This explicitly disavows the dubious operation of changing a curve's currency.

In practice discounting and forecasting are often tied together: *e.g.*, we may write

$$F(t_x) = \delta_t^{-1}\left(\frac{Z(t_s)}{Z(t_m)} - 1\right)$$

where t_x, t_s, t_m are fixing, start and maturity times respectively, δ_t is a daycount fraction and Z is a zero-coupon bond price from a discount curve. This gives forecasts – *e.g.*, of Libor rates – as a function of discounts, for some subset of possible yield curves. The discount curve which yields on-market forecasts of Libor is called "Libor flat."

12.2.1 *Libor*

For many years, Libor flat discounting was practically a standard, and dealers sought small price improvements by considering "funding adjustments", perturbations of the discount curve away from the Libor curve. This practice turned out to depend on an unjustified faith in orderly market processes, in precision in the polled fixing of Libor rates, and in reliably available and homogenous funding for banks.

To construct a curve adequate for trading, we must take into account several facts about the world:

- Libor is the averaged output of a polling process, not a market-clearing interest rate for any instrument.
- Term borrowing rates have fine structure from adjusting maturity to be a good business day; Libor rates do not.
- The cost of borrowing depends strongly on what collateral can be provided.
- The currency of an asset is an important factor in its desirability as collateral.

The most liquid and frequently traded interest-rate instruments are still swaps, though these are now centrally cleared and collateralized. Thus the fundamental components of a yield curve are a discount curve for standard collateral (government bonds acceptable to the swaps clearinghouse), and a forecast curve of Libor rates, at least for the standard Libor tenor. We need to decide how to construct this forecast curve, and how to fit the two curves to available market data.

While the standard procedure for obtaining forecasts from a given discount curve is not entirely justified for Libor rates, it at least exists, whereas there is no corresponding way to recover discounts from a forecast curve. Thus we will treat some discount curve as the fundamental part of our parametrization. Our preference is to choose the collateralized discount curve for this role, and treat Libor-flat discounting as a perturbation thereof, but the converse is also defensible.

Let us turn for a moment to the problem of obtaining Libor forecasts from a discount curve. It is now widely known that any price obtained with Libor-flat discounting is *ipso facto* wrong; thus we need not worry about adhering to the defunct market standard and suffering its pernicious effects. For instance, 3-month USD Libor fixing on a Wednesday in March has a nominal start date on Friday; 3 months forward from that is a Saturday in June, so the adjusted maturity date would roll to the following Monday. In an upward-sloping yield environment, this would lead Wednesday's predicted Libor fixing to be substantially above Tuesday's. This effect, an artifact of mistaken assumptions about the nature of Libor, is reliably absent from reality.

By omitting details such as maturity adjustment from the calculation process, we arrive at results which are both more robust and more realistic. Our preference is to ignore the fixing delay, and use a nominal 365-day year to convert the quoted tenor to an offset in days. Thus we arrive at the code:

```
————————————————— YCImp.cpp ——————————
double LiborForecastFromDiscounts
    (const DiscountCurve_& dc,
     const Date_& fix_date,
     int tenor_months,
     int tenor_weeks,
     const DayBasis_& daycount)
{
    auto end = fix_date.AddDays
        ((365 * tenor_months) / 12 + 7 * tenor_weeks);
    const double df = dc(fix_date, end);
    return (1.0 / df - 1.0) / daycount(fix_date, end, nullptr);
}
```

This code may differ from any market standard, but with no observable consequences, because no observable discount curve corresponds to market Libor rates. We are taking advantage of the fact that the Libor-flat curve is a computational fiction, and choosing it to have the properties we desire.

Given a discount curve, it is now simple to create a dedicated forecast curve based on those discounts, for a given Libor rate. Note that each Libor rate has its own forecast curve, because each will need to be fitted separately to available market data. Any attempt to deduce the expected results of one Libor rate poll from those of a different poll is a delusion.

12.2.2 *Parametrization*

A good parametrization for discount curves is piecewise linear forward rates,[1] which

- Are a superset of both commonly used bootstrap methods (linear zero rates and linear log discounts);
- Allow a smooth structure of term rates without the overshoot problems plaguing splines;
- Support the introduction of additional degrees of freedom to explicitly address the underdetermined nature of a curve fit; see Sec. 7.4.

This immediately leads us to a basic implementation of the discount curve:

```
                          YCImp.cpp
   class PWLF_ : public DiscountCurve_
   {
       Handle_<DiscountCurve_> base_;
       PiecewiseLinear_ fwds_;
5      PWLF_(const String_& nm, fwds_, base_) : DiscountCurve_(nm) {}

       double operator()(const Date_& from, const Date_& to)
       const override
       {
10         return exp(fwds_.IntegralTo(from) - fwds_.IntegralTo(to))
                  * (base_ ? (*base_)(from, to) : 1.0);
       }
       void Write(Archive::Store_& dst) const override;
   };
```

This class supports both a fundamental curve (*e.g.*, that for collateralized discounting) and a perturbation to an existing base curve (*e.g.*, the curve underpinning Libor-flat forecasting).

The `base_` has further applications, such as inserting turn-of-year costs and other manually configurable features, then fitting a smooth perturbation to match market prices.

---

[1] Pioneered by Tom Coleman at TMG.

12.2.3   *Fitting*

A full yield curve fit still faces a substantial obstacle: the quoted instruments involve multiple forecast and discount curves with no clear ordering. We need collateralized discounting rates to value a Libor swap, but the only liquid quotes for those rates (except at the very front end) are in the form of Libor-OIS basis swaps, or their non-USD equivalents. Thus it is generally necessary to fit at least two curves simultaneously; in currencies like EUR and GBP, where swaps reference 6-month Libor but the futures market settles to 3-month rates, the problem is even worse.

Fortunately, the fitting problem is nearly linear, and underdetermined methods (see Sec. 7.4) work very efficiently.

It is still difficult to build an extremely fast curve, because of the daily compounding convention of OIS and similar swaps; but bear in mind that basis swap quotes are not updated rapidly in the market, so our most frequent task is only to recalibrate the overall level of rates with fixed spreads. The careful student of underdetermined search will have noticed that it allows us to specify an arbitrary initial guess, and look for a nearby solution point. We can use this to choose the most recent full curve *as the starting point* for a search which will match the most liquid swaps and futures, while keeping forward rate spreads fixed (so spot spreads will move minimally). This "fast/slow" paradigm is now widespread.

The implementation of these concepts, including user specification of a customizable *build stack* which controls the available degrees of freedom and couplings between them, is quite elegant but outside the scope of this volume.

12.3   Build Instruments

Fitting the forward curve to a set of market quotes for standard instruments (the *build instruments*) is a fundamental task. The quotes themselves are generally also available as index fixings; thus we have already created a naming scheme for them, in Sec. 9.1. However, the scheme there cannot take advantage of prior information – that this is an interest rate instrument in a specified currency – and thus yields overly verbose names.

We address this by supporting long instrument names which are the full index names with the IR[*ccy*] dropped from the front: *e.g.*,

SWAP:LIBOR:5Y. Even these may be considered too long, and we also support an independent set of short names, like 5Y and Sep14, which do not map directly to index names.

Ideally the index parsing and curve instrument parsing would share implementation, but the instrument parser is sufficiently simple that we can relax this constraint. It is sufficient that the same schedule-generation function, which contains all the fragile code, be used for both rate indices and curve instruments.

An instrument supports curve fitting by computing an implied rate, given a candidate curve. Because curve fitting is so frequent, it is worthwhile to optimize this computation; we do this by separating the rate computer from the instrument itself.

```
—————————— YCInstrument.h ——————————
class YcInstrument_ : noncopyable
{
public:
    virtual ~YcInstrument_();
5   virtual String_ Name() const = 0;
    virtual pair<Date_, Date_> TimeSpan() const = 0;

    struct Rate_ : noncopyable
    {
10      virtual ~Rate_();
        virtual double operator()
            (const YieldCurve_& yc)
        const = 0;
    };
15  virtual Handle_<Rate_> Precompute
        (const Handle_<YcInstrument_>& self,
        const Handle_<YieldCurve_>& funding_yc)
    const = 0;
};
```

The Precompute function produces the object which will actually compute the rate – all schedule generation, which cannot change across curve scenarios, can be done at this stage. The function's two inputs each correspond to an additional possible optimization. The input self will always satisfy self.get() == this; thus instruments which need no optimization can derive from Rate_ and return self from Precompute. The second input, if non-null, tells the instrument that funding spreads will not vary during the fitting process, so they can be computed once and for all at this stage. This accelerates the most common kinds of curve builds.[2]

12.4 Dividend

In equity models with deterministic dividend yields, we create a "dividend curve" which may be either input directly or calibrated to market equity forward prices. This introduces no new challenges, but a couple of points are worth noting:

- Though the dividend curve may have the same implementation as a rate or spread curve, it should never be implemented in terms of such a curve. This gratuitously complicates the construction and especially the perturbation of such curves, making sensitivity computations more difficult and restricting implementation changes to either one. Shared functionality should, as always, be factored out into lower-level code.
- Similarly, a two-currency model can be converted into an equity model by converting the equity dividend curve to a yield curve for a fake foreign currency; this may be necessary in the short term but is an obstacle to the longer-term goal of creating component-based models capable of handling both FX and equity.

Production models for equity derivatives should support cash dividends, and even for equity indices the assumption of proportional dividends is undesirable.

12.5 Hazard

To price payouts contingent on default or non-default of some *reference credit*, we must extract the *default intensity* from market instruments. These are generally credit default swaps (*CDS*), or occasionally bonds with issuer risk. We are interested in $Q(T)$, the survival probability to time T in the discount-adjusted measure; and in the default intensity or *hazard rate*, $\lambda(T) \equiv -Q'(T)/Q(T)$. Because we define Q in the appropriate measure, it can immediately be used to compute CDS and bond prices.

Recall from Sec. 8.5.3 that a CDS is described by a fixed coupon *premium leg* with payments c_i at times t_i, $i = 1 \ldots n$; and a *protection leg* of

---

[2]We can go further and supply a partial yield curve, built up to the previous instrument end date, to `Precompute`; but this chains us to bootstrapped curves.

default protection on a notional $N(t)$. For a standard CDS, $N(t) = N_0 \mathbf{1}_{t < t_n}$ is constant up to the CDS maturity, and the value of the protection leg is

$$N_0 \int_0^{t_n} Q(t)Z(t)\lambda(t)dt,$$

where $Z(t)$ is a riskless cash discount factor. The premium leg value, complicated by the payment of accrued interest in the event of default, is

$$\sum_i c_i \left(Q(t_i)Z(t_i) + \int_{t_{i-1}}^{t_i} \frac{u - t_{i-1}}{t_i - t_{i-1}} Q(u)Z(u)\lambda(u)du \right).$$

These integrals are a convenient notation, but we must be aware that they are continuous-time approximations to discontinuous processes; discounting of cash flows is an intrinsically daily event, and default is nearly as discrete.

For valuation purposes, we might choose a coarser discretization to improve performance. It is tempting to use knowledge of the form of λ or even of Z to press this still further: a few exponential integrals, plus some Taylor expansions to handle edge cases, give a feeling of accomplishment. But the more we consider the issues of flexibility and maintenance – driven by desire to build hazard curves that will continue working flawlessly, even as other developers change the fit algorithm or the character of the yield curve – the more merit appears in the brute-force approach.

There is one valuable optimization for the fitting of hazard curves. We will evaluate CDS rates for many candidate solutions, each time with the same $Z(t)$; and, since $\lambda(t)dt = -dQ(t)$ we can eliminate λ in favor of Q everywhere and write the value of a CDS as a linear combination of Q-factors. This precomputed representation can be formed independently of the parametrization chosen for Q or for Z. For a specific fit method, where the parametrization is known, we can optimize this representation further while retaining very high accuracy.

Chapter 13

Models

To provide numerical pricing, models and their SDE's (see Sec. 10.2.1) must implement the protocols of Ch. 10. Much of this functionality can be provided by the base SDE_ class;[1] or, if the implementation does not require access to private members, in supporting free functions. The latter approach, which separates functionality from the requirement for a particular sort of this pointer, is preferred.

Rather than plunge directly into these implementations, we explore the implementation of a specific simple model. In the process we will develop a mix of specific and general functionality; the latter should be easily recognizable, and will mainly be collected in ModelImp.cpp.

13.1 Vasicek-Hull-White

The Vasicek model is usually specified by giving an SDE for the short rate, r:

$$dr = \kappa(\mu - r)dt + \sigma dW$$

where κ and σ may be time-dependent, and μ must be strongly time-dependent to reproduce zero-coupon bond prices.

This representation is suboptimal, for two reasons. First, the behavior of μ adds complexity to what is at heart a very simple model; the implemention must concern itself with derivatives of forward rates, which cannot be considered financially meaningful. The second problem is that, by introducing the idea of mean reversion of a short rate, we are exaggerating the role of the short rate, and diverting attention from real tradable quantities.

---

[1] Actually, we prefer to make SDE_ a pure interface and place shared functionality in SDEImp_, from which concrete SDE's are derived.

The mean reversion rate κ determines the values of options, but in a very non-linear way.

These drawbacks can be eliminated, if we instead write

$$r = E[r] + HS, \qquad dS = gdW$$

where g and H are deterministic functions of time, and the expectation is of course in the risk-neutral measure. We then find $\sigma(t) = g(t)H(t)$, $\kappa(t) = -H'(t)/H(t)$. This representation has several advantages:

- Volatilities are linear in the product gH, and option prices nearly so, expediting calibration.
- The HJM volatility $\sigma(t,T) = g(t)H(T)$ is transparently determined.
- Any one-state-variable model is limited to a single mode of yield curve motion; H is exactly this mode.
- The relation of the term structure of g and H to swaption prices is instantly comprehensible.

Thus we end up with a cleaner and marginally more efficient implementation, coupled with a notation which clearly relates the model parameters to option prices.

All the dynamics of the model will depend on a few integrals:

- $\int_{t_-}^{t_+} g^2(u)du$, the variance of S over an interval;
- $B(t,T) \equiv \int_t^T H(u)du$, the sensitivity of bond prices to S;
- $-\int_0^T g^2(t)B(t,T)dt$, which is $\tilde{E}[S]$ in the discount-adjusted measure;
- $\int_{T_-}^{T_+} g^2(t)B^2(t,T_+)dt$, the variance of the innovation in discount factors.

The last integral is required only for Monte Carlo simulation in the risk-neutral measure, which we can avoid; see Sec. 13.1.2.

13.1.1 *Parametrization*

Now, how shall we parametrize g and H? Few-parameter forms should be rejected immediately: the model is already perfectly tractable, and we cannot afford to further reduce its already meager market-fitting capability. In addition, once we have a flexible nonparametric form, we can closely approach any desired parametrization with a special-purpose constructor.

A possible nonparametric form is to make both g and H piecewise-constant functions of time. A generation of quants, trained to believe they are adding value by analytically evaluating complicated integrals, may hold such a simplistic form in contempt; but there is in fact no business justification for anything more complex. Even the smooth exponential decay of H (and growth of g) which leads to a stationary model can be quite well approximated with this method.

Given this parametrization, we store for each interval the values of g and H, and the starting values of the integrals above, accumulated from zero. Since these are used internally by the model, not displayed to calling functions, we place them in **namespace VHWImp**:[2]

```
——————————————— VHWImp.h ———————————————
namespace VHWImp
{
    struct Piece_
    {
        double g_;
        double h_;
        double B_;
        double varS_;
        double covDS_;
        double varD_;
    };

    class Vol_ : public Storable_
    {
        map<DateTime_, Piece_> vol_;
        void Write(Archive::Store_& dst) const override;
    public:
        Vol_(vol_, const String_& name = String_());

        void Integrate
            (const DateTime_& from,
             const DateTime_& to,
             double* b,
             double* var_s,
             double* cov_d_s = nullptr,
             double* var_d = nullptr)
        const;

        Handle_<PiecewiseConstant_> G() const;
        Handle_<PiecewiseConstant_> H() const;
        DateTime_ VolStart() const{ return vol_.begin()->first; }
    };

    Vol_* NewVol
        (const DateTime_& vol_start,
         const PiecewiseConstant_& g,
```

—————————————————————————————

[2] The last two quantities named are a covariance and variance of $\log(D)$, not of D itself, but there is no need to jam this information into their names.

```
        const PiecewiseConstant_& h);
}
```

The query functions G and H allocate and populate the return structure, so they are not very efficient; we do not mind this because they are not used in pricing.

NewVol lets us construct the volatility from separate PWC_ structures, possibly with different time-dependence; for example, *g* might be constant if we are only considering options with a single common expiry.

This Vol_ is itself storable; we will write it by writing *g* and *H*, and compute the derived quantities when reconstituting a stored value.

VHWImp.cpp
```
  /*IF-----------------------------------------------------------
  storable VHW_Vol
      Volatility for VHW model
  version 1
5 manual
  &members
  name is ?string
  g is handle PiecewiseConstant
      The (time-dependent) volatility of the state S
10 H is handle PiecewiseConstant
      The (time-dependent) sensitivity of F(T) to S
  -IF----------------------------------------------------------*/
  // ...
  #include "MG_VHW_Vol_Write.inc"
15 #include "MG_VHW_Vol_Read.inc"

  VHWImp::Vol_* VHW_Vol_v1::Reader_::Build() const
  {
      Vector_<DateTime_> knots
20        = Unique(Join(g_->knotDates_, H_->knotDates_));
      map<DateTime_, Piece_> vals;
      for (auto t : knots)
      {
          vals[t].g_ = PWC::F(*g_, t);
25        vals[t].h_ = PWC::F(*H_, t);
      }
      return new VHWImp::Vol_(vals, name_);
  }
```

13.1.2   Model Contents

A yield curve, plus a Vol_, is the complete specification of the Vasicek-Hull-White model. Additionally, the model unambiguously obeys a known

SDE, whose parameters do not depend on the trade underlying (see the discussion in Ch. 10). Thus we implement both model interfaces:

```
                        ─────── VHW.cpp ───────
   class VHW_ : public Model_, public SDEImp_
   {
       Handle_<YieldCurve_> yc_;
       Handle_<Vol_> vol_;
5
       void Write(Archive::Store_& dst) const
       {
         VHW::XWrite(dst, name_, yc_, vol_);
       }
10 public:
       VHW_(const String_& name,
           const Handle_<YieldCurve_>& yc,
           const Vol_& vol);
```

Note that this is in a source file; there is no need to gratuitously parade the class definition in a header. As always, we will use factory functions to create actual instances.

13.2   Interface to Numerical Pricing

We connect the model to our numerical pricers using the model's own state variables, with no transformation. Monte Carlo simulation is, of course, agnostic to this choice, since it does not manipulate the state variables but only passes them to Stepper_s and asset models. By displaying a state variable with highly time-dependent volatility[3] to a PDE, we are committing to having our PDE solver support a time-dependent grid size; this is not difficult and adds value for many different models.

A stepper must describe the evolution of S over an arbitrary interval from t_- to t_+, but the no-arbitrage constraints of the model can be satisfied in several ways.

- We can evolve S in the pure risk-neutral measure, so that the stochastic discount factor D also has an innovation over the interval. This requires two random numbers, since $\log(D)$ is not perfectly correlated with S.
- We can suppress the idiosyncratic innovation in D, using $E[D; S]$ instead; more precisely, $E[D_+; D_-, S_-, S_+]$ where $D_\pm, S_\pm$ refer to quantities at time $t_\pm$. This can be thought of as a path integral, over the path of S between the event times.

---

[3]Recall that $g(t)$ may be exponentially growing.

- We can suppress the innovation in D altogether, using instead $E[D; D_-, S_-]$. We then adjust the drift of S for the implied change of measure. This constructs the *jumping numeraire*.

We want our stepper to consume only a single Gaussian deviate per step:

```cpp
                              VHW.cpp
class VHWStep_ : public ModelStepper_
{
    double sqrtDt_;
    double muS_, sigmaS_;    // de-annualized!
5   double a_, bMinus_, bPlus_;
public:
    MonteCarlo::Workspace_* NewWorkspace
        (const MonteCarlo::record_t&)
        const override  { return 0; }
10  int NumGaussians() const override { return 1; }

    void Step
        (Vector_<>::const_iterator iid,
         Vector_<>* state,
15       MonteCarlo::Workspace_*, Random_*,
         double* rolling_df,
         Vector_<Handle_<DefaultEvent_>>*)
        const override
        {
20      const double sMinus = state->front();
        state->front() += muS_ + sigmaS_ * *iid;
        *rolling_df *= exp(a_ - bMinus_ * sMinus
                 - bPlus_ * state->front());
        }
}
```

Obviously we require $B_+ + B_- = B(T_-, T_+)$. If $B_+ = 0$, this implementation reduces to the jumping numeraire. We prefer

$$B_+ = \int_{T_-}^{T_+} g^2(t)B(t, T_+)dt / \int_{T_-}^{T_+} g^2(t)dt,$$

i.e., we reproduce the covariance of S with $\log D$; thus we can set $\mu_S = 0$. Now the joint distribution of $S_\pm$ is known, and we derive a to fit the yield curve, finding

$$a = \log\left(\frac{Z_+}{Z_-}\right) + (B_- + B_+)\tilde{E}[S_-] - \frac{B_-(B_- + 2B_+)}{2}\text{Var}[S_-] - \frac{B_+^2}{2}\text{Var}[S_+],$$

where $Z_\pm$ denotes the discount factor to $T_\pm$ and $\tilde{E}$ is an expectation in the discount-adjusted measure to time T_-. Note that the calculation of $\tilde{E}[S]$ is already supplied by the parametrization described in Sec. 13.1.1.

Next we proceed to form the quantities needed for (backward induction) PDE pricing. We have postulated that S is a martingale, so we need no advection term; the diffusion term is `sigmaS_ / sqrtDt_`.[4] Only the discounting is S-dependent, with the rate

$$(B_- + B_+)S - \frac{a}{\Delta_t}.$$

Thus only the discounting function requires its own implementation class.

```
───────────────────────── VHW.cpp ─────────────────────────
     PDE::ScalarCoeff_* DiscountCoeff() const override
     {
         struct Mine_ : PDE::ScalarCoeff_
         {
 5           x_dep_t xDep_;
             double c0_, c1_;
             Mine_(c0_, c1_) { xDep_.set(0); }

             void Value
10           (const Vector_<>& x, double* value) const override
             {
                 assert(x.size() == 1);    // just S
                 *value = c0_ + c1_ * x[0];
             }
15           x_dep_t XDependence() const { return xDep_; }
         };
         const double dt = Square(sqrtDt_);
         return new Mine_(a_ / dt, (bPlus_ + bMinus_) / dt);
     }
20   PDE::VectorCoeff_* AdvectionCoeff() const override
     {
         return PDE::NewConstCoeff(Vector::V1(0.0));
     }
     PDE::MatrixCoeff_* DiffusionCoeff() const override
25   {
         return PDE::NewConstCoeff(SquareMatrix::M1x1
             (0.5 * Square(sigmaS_ / sqrtDt_)));
     }
     };
```

We do not use the `StepAccumulator_` in creating steppers, but we still must provide envelope information for PDE solvers.[5]

```
───────────────────────── VHW.cpp ─────────────────────────
struct VHWAccumulator_ : StepAccumulator_
{
```

---

[4] If `Step` is used at all, it will be called far more frequently than the PDE coefficient calculators; thus we favor it in our optimization.

[5] Here we construct the envelope in the discount-adjusted measure. It might be considered safer to give the upper bound in the risk-neutral measure (remember $\tilde{E}[S] \leq 0$).

```
      DateTime_ volStart_;
      const VHWImp::Vol_& vols_;
 5    Vector_<pair<double, double>> Envelope
          (const DateTime_& t,
           double num_sigma)
      const override
      {
10        double varS, eS;
          vols_.Integrate(volStart_, t, 0, &varS, &eS, 0);
          const double range = sqrt(varS) * num_sigma;
          return Vector::V1(make_pair(eS - range, eS + range));
      }
15  };
```

13.3   Interface to Valuation Requests

Upon receipt of a completed value request, we must create an asset model
which will honor those requests by computing the desired market quantities,
as a function of the model state S. The SDE_ class will provide a template
method to create an updater, based on calls to the model to produce a
value for a single index.

Many of the steps in this process can share the same implementation
across all model types, though they are not part of the SDE_ interface. We
exploit this by creating a derived SDEImp_, which provides this common
skeleton for all models.

13.3.1   *Updating One Realization*

In the VHW model we assume that changes in the state drive all discounts,
including those implied from Libor rates, in the same way; *i.e.*, funding
adjustments are deterministic. A Libor rate starting from t_d can used to
construct a discount factor, equal to $1 + \delta L$ where δ is the daycount fraction,
over the interval from t_d to the Libor maturity t_m. Thus we have:

$$1 + \delta L = A e^{B(t_d, t_m)S},$$

where A is determined by the necessity of repricing Libor in the discount-
adjusted measure to t_m:

$$1 + \delta F = A \tilde{E}_{t_m} \left[e^{B(t_d, t_m)S} \right]$$

where F is the forward Libor rate from the yield curve. Thus we know our final code must rely on a supporting function:[6]

```
———————————————————— VHW.cpp ————————————————————
SDEImp::UpdateOne_ * NewLiborUpdater
    (_ENV, const YieldCurve_& yc,
     const VHWImp::Vol_& vol,
     const DateTime_& event_time,
     const Date_& start_date,
     const PeriodLength_& tenor)
{
    struct Mine_ : SDEImp::UpdateOne_
    {
        double A_, B_, dct_;
        Mine_(A_, B_, dct_) {}
        double operator()
            (const Vector_<>::const_iterator& state,
             const UpdateToken_&)
        const override
        {
            return (A_ * exp(B_ * state[0]) - 1.0) / dct_;
        }
    };

    const double F = yc.FwdLibor(tenor, start_date);
    const Date_ maturity = Date::NominalMaturity
        (start_date, tenor, yc.ccy_);
    const double dct = Ccy::Conventions::LiborDayBasis()(yc.ccy_)
        (start_date, maturity, nullptr);
    double B, ES, varS;
    vol.Integrate(DateTime_(start_date),
        DateTime_(maturity), &B, nullptr);
    vol.Integrate(vol.VolStart(), event_time, nullptr, &varS, &ES);
    const double A = (1 + dct * F) * exp(-B * (ES + B * VarS / 2));
    return new Mine_(A, B, dct);
}
```

This function will be called from a member function of `VHW_`, such as `VHW_::NewUpdater`, when a trade requests a Libor rate. The value of `A` is chosen to ensure the correct expectation of the rate in the discount-adjusted measure.[7]

The extra `UpdateToken_` argument in the signature of `UpdateOne_` is not needed here, because our model has perfect self-knowledge: to compute a Libor, we need the `state` and nothing else. However, when we proceed to

---

[6]The only subtle feature of this function is the choice of arguments to `operator()`, which we are about to explain.

[7]Those familiar with short-rate Hull-White models might appreciate the conspicuous lack of nested integrals in this computation.

component-based models (see Sec. 13.7) we must permit the components
to *make requests of each other* to separate their implementations. For
instance, an equity component will fulfill a request for a forward equity price
by in turn requesting a discount factor – thus avoiding the need for it to
know the details of how discounts are computed from the interest-rate state.

Similarly, we pass the `state` as an iterator rather than a whole vector,
because the parent `Vector_<>` might be the state of some larger hybrid
model. Our component need not know its parent's layout; it is responsible
only for the sub-vector containing its own state.

These two considerations determine the interface of the general updater,
of which `NewLiborUpdater` returns an instance:

```
———————————— SDEImp.h ————————————
    namespace SDEImp
    {
        class UpdateOne_ : noncopyable
        {
5       public:
            virtual ~UpdateOne_();
            virtual double operator()
                (const Vector_<>::const_iterator& state,
                 const UpdateToken_& prior)
10          const = 0;
        };
    }
```

13.3.2   *Updating for One Node*

Individual updates are weakly ordered by their possible dependency rela-
tionships. The requests which the model receives from the trade are ordered
by their `IndexKey_`, but this is not relevant to the model. We must produce updaters which are called in the correct sequence, so that the needed
`prior` information is always available for a given update. The dependency
tracking for this is handled inside `SDEImp_`, so the interface to the asset can
be at the highest level:

```
———————————— SDE.h ————————————
    namespace SDEImp
    {
        class Update_ : noncopyable
        {
5       public:
            virtual ~Update_();
            virtual void operator()
                (Vector_<>* vals,
                 const Vector_<>& state,
```

```
10        const UpdateToken_& pass)
        const = 0;
    };
}
```

The `operator()` will be called with `pass.begin_ == vals->begin()`; thus the `Update_` can populate `vals` while also allowing `UpdateOne_` functions to access it as their `prior`. The `SDE_` is responsible for ensuring that this unsafe access is always in the correct order. This responsibility, and the presence of the entire `Vector_<>` of `state`, make this top-level updater the province of the parent `SDE_`, not any of its components.

Obviously there is an `Update_` for each event time; creation of these requires non-`const` access to the `ValueRequest_` being satisfied. We provide `SDEImp_` with the member function:

```
                        ─── SDE.h ───
    SDEImp::Update_* NewUpdate
        (_ENV, ValueRequestImp_& requests,
         const DateTime_& event_time)
    const;
```

`ValueRequestImp_`, which is the base type of all concrete `ValueRequest_s`, is discussed in Sec. 10.8.2. Here we will see the need for its `AtTime` query.

To implement `Update_` with a collection of `UpdateOne_`, we need to store the single updates and the destinations for their outputs.

```
                    ─── SDEImp.cpp ───
struct UpdateImp_ : SDEImp::Update_
{
    Vector_<pair<Valuation::address_t,
        Handle_<SDEImp::UpdateOne_>>> updates_;

5
    void operator()
        (Vector_<>* vals,
         const Vector_<>& state,
         const UpdateToken_& pass)
10  const override
    {
        assert(&pass[0] == &vals->front());
        for (auto u : updates_)
            (*vals)[u.first] = (*u.second)(state.begin(), pass);
15  }
};
```

This will only work if we ensure that the `updates_` are executed in order. To track dependencies, the `SDEImp_` must mediate access to the

`ValueRequest_`; thus we create another object to pass to derived `SDE_s` making single-index evaluators.

```
─────────────────────── SDEImp.h ───────────────────────
class RequestAtTime_
{
public:
      const DateTime_ t_;
5     Valuation::address_t operator()(const Index_& index);
private:
      ValueRequestImp_& request_;
      map<IndexKey_, Valuation::address_t> all_;
      bool stale_;
10    set<IndexKey_> done_;
      Vector_<Handle_<Index_>> wait_;    // priors

      friend class SDEImp_;
      typedef Handle_<SDEImp::UpdateOne_> update_t;
15    RequestAtTime_(ValueRequestImp_& r, t_)
            : request_(r), all_(r.AtTime(t_)), stale_(false) {}

      void Refresh();
      void Push
20       (Vector_<pair<Valuation::address_t, update_t>>* updates,
          const IndexKey_& index,
          const update_t& update);
};
```

This `operator()` will obtain a promised fixing location from its `request_`, and will also note whether the request is a new one (whose update must therefore precede the update being formed). The `stale_` flag indicates whether `all_` needs updating – *i.e.*, whether more index requests have been made since `AtTime` was called.

```
─────────────────────── SDE.h ───────────────────────
Valuation::address_t RequestAtTime_::operator()
      (const Index_& index)
{
      IndexKey_ key(Index::Clone(index));
5     if (!done_.count(key))
          wait_.push_back(key.val_);
      if (!all_.count(key))
          stale_ = true;
      return request_.Fixing(t_, index);
10 }
```

We store two containers based on the index's key;[8] the set done_ of indices for which we have already produced and stored an UpdateOne_, and the map all_ of indices for which promises have been made. If the key is not in done_, then we need to add it to the stack of continuations (see below), and later ensure that its updater is run before the updater for whatever index requested it. If the key is not in all_, then this is a truly new request – not requested directly by the trade, nor as part of any previous index setup – and we will eventually need to refresh all_.

We store the dependencies of some updates on others as *continuations* which associate a deferred task with those that must precede it.

```
———————————— SDEimp.cpp ————————————
struct Continuation_
{
    Handle_<Index_> index_;
    Handle_<SDEImp::UpdateOne_> update_;
    Vector_<Handle_<Index_>> wait_;
    Continuation_(index_, update_, wait_) {}
};
```

Now we can finally implement NewUpdate:

```
———————————— SDEimp.cpp ————————————
SDEImp::Update_* SDEImp_::NewUpdate
    (_ENV, ValueRequestImp_& requests,
     const DateTime_& event_time)
const
{
    unique_ptr<UpdateImp_> retval(new UpdateImp_);
    Stack_<Continuation_> tbc;
    RequestAtTime_ toDo(requests, event_time);
    for (;;)
    {
        Handle_<SDEImp::UpdateOne_> update;
        Handle_<Index_> next;
        if (tbc.empty())
        {   // no continuations
            toDo.Refresh();
            if (toDo.all_.size() == toDo.done_.size())
                break;
            next = FindAnother(toDo.all_, toDo.done_);
        }
        else if (tbc.top().wait_.empty())
        {   // now ready for this one
            next = tbc.top().index_;
            update = tbc.top().update_;
```

[8]The call to Index::Clone here is somewhat inefficient, and could be improved by ensuring the index is already in a Handle_.

```
                        tbc.pop();
25      }
        else
        {   // still clearing the decks
            next = tbc.top().wait_.back();
            tbc.top().wait_.pop_back();
30      }

        // get an update, and see whether we have to wait
        assert(toDo.wait_.empty());
        if (update.Empty())
35              update.reset(NewUpdateOne(_env, toDo, *next));

        if (toDo.wait_.empty())
        {   // we can do it now
            toDo.Push(&retval->updates_, next, update);
40      }
        else
        {   // put it on hold, do others first
            tbc.push(Continuation_(next, update, toDo.wait_));
            assert(!IsCircular(tbc.Peek()));
45          toDo.wait_.clear();
        }
    }
    return retval.release();
}
```

This code is complex, but its general outline should be clear enough. We start by choosing an index in `all_` but not `done_` – the routine `FindAnother` accomplishes this. We create an `UpdateOne_` for it, with `toDo` to tell us what other updates it depends on. Since the call to `Push` updates `toDo.done_` and `retval->updates_` are together, they are reliably synchronized.

If the new update depends on updates which are not `done_`, then we postpone it – pushing in onto a stack of `Continuations_` on top of those updates which depend on it – while we work through those dependencies. When an index is taken out of `tbc.back().wait_`, it next appears either in `done_` or as a new entry in `tbc` awaiting its own turn; either way, it is sure to be processed before `tbc.back().index_`, whose update depends on it. Thus no update is pushed to `retval` (by the call to `toDo.Push`) until all the updates it depends on have already been pushed.

This routine will fail utterly – with a stack overflow – if there is a circular dependency among the updates. We can detect this by testing that the `index_` entries of each stored `Continuation_` are unique; this is the function of `IsCircular` in the displayed code.[9] This must arise from a grievous coding error, so an `assert` is appropriate.

This might seem recondite, but it is crucial for robust functioning even of simple models. For instance, Sec. 13.3.5 shows an `UpdateOne_` which computes a swap rate from Libor forwards; this relies on the Libor rates' all being updated before we attempt to update the swap rate. Our careful implementation of `NewUpdate` ensures that this will be the case – *simply by requesting the Libor rates* when computing the swap updater, the `SDE_` signals the dependency needed to enforce the correct calculation order. This savings in code and in programmer effort is well worth a few dozen lines of abstruse code.[10]

13.3.3 *Index Paths*

In the absence of index path requests, an `Update_` for each of the trade's event times is sufficient to form an `Asset_`. To form index paths, we need to add to `SDEImp_` a new member:

```
────────────────── SDEImp.h ──────────────────
IndexPath_* NewIndexPath
    (const Index_& index,
     const Vector_<DateTime_>& index_times,
     ValueRequest_& request);
```

The input `index_times` are the trade's event times, excluding those beyond the last date on which the `IndexPath_` is used.[11] The path will naturally depend on the fixings at event dates, which we will obtain from the `ValueRequest_`; thus the index-path updates are *created before* but *executed after* the fixing updates.

Information about path behaviour between event dates – *e.g.*, volatilities for a Brownian bridge – will generally not require cross-index updates. Composite indices can compute such quantities only if the model describes the joint behavior of indices between event dates; this is beyond the scope of this volume.

---

[9] Actually, `IsCircular` must examine the whole `stack` of continuations, so it will be grossly inefficient unless we change `tbc` to a vector.

[10] An alternative is to embed in code a hierarchy of index types with dependencies all in one direction; *e.g.*, ensure that every Libor rate is computed before any swap rate. This is verbose and not extensible; the code shown here is the result of an effort to transcend it.

[11] There is a very slight inefficiency here, in that we require the construction of the vector of index_times rather than just passing iterators.

13.3.4 *Efficiency*

When we are working through the continuation stack in NewUpdate, we know we have not exhausted the requested indices; thus we Refresh only when the stack is empty, when we ask the request for more tasks. For simple models without the possibility of dependencies between requests, tbc will always be empty and the flow of control inside the loop reduces to:

```
     if (toDo.all_.size() == toDo.done_.size())
         break;
     IndexKey_ next = FindAnother(toDo.all_, toDo.done_);
     Handle_<SDEImp_::UpdateOne_> update
5            (NewUpdate(toDo, *next));
     toDo.Push(&retval->updates_, next, update);
```

The completeness check and NewUpdate take $O(1)$ time, regardless of the number of indices being updated; and Push takes $O(\log n)$ time because of the call to set::insert. A naive implementation of FindAnother would walk through all_ and done_ together, taking $O(n)$ time for each call (thus $O(n^2)$ total time). We can avoid this by writing:

```
————————————————— SDEimp.cpp ———————————————
Handle_<Index_> FindAnother
    (const map<IndexKey_, Valuation::address_t>& all,
     const set<IndexKey_>& done)
{
5    assert(all.size() > done.size());
     auto pa = done.empty()
         ? all.begin()
         : all.upper_bound(*done.rbegin());
     if (pa == all.end()) // fast method failed
10   {
         pa = all.begin();
         auto pd = done.begin();
         while (pd != done.end() && pa->first == *pd)
             ++pa, ++pd;
15   }
     return pa->first.val_;
}
```

which first tries to find something in all after the last element of done (taking $O(\log n)$ time), before falling back on the linear search. This could be further optimized by providing the last returned value as a hint, and starting the linear search there rather than always at all.begin().

Another optimization has to do with code caching. If several updates – *e.g.*, of equities within a basket – are evaluated by the same code, then we can reduce cache faults and improve performance by grouping those updates together. This requires a functor to inspect an index and determine whether its evaluation should be brought forward:

```
—————————————— SDEImp.h ——————————————
struct IndexIsSimilar_ : noncopyable
{
    virtual bool operator()(const Index_& index) const = 0;
};
```

Then we add

- A local variable, `scoped_ptr<IndexIsSimilar_> sim`, alongside `monitor` and `tbc` above;
- An additional argument, `scoped_ptr<IndexIsSimilar_>* sim`, to the single-index version of `NewUpdate` – we will send `&sim` to this routine;
- An additional argument, `const IndexIsSimilar_* sim`, to `Find Another` – we will send `sim.get()`.

Each call to `NewUpdate` has the option to reset `sim` to a new test function; when a non-NULL test function is provided, `FindAnother` must attempt to return a preferred index, but also account for the possibility that no such index exists. We can accomplish this by creating a circular iterator which loops around `all`, starting from the initial `pa` above; and saving a fallback value to be returned if `(*sim)(x)` is `false` for all keys x. This code can no longer be expected to run in $O(\log n)$ time, so it can make the setup phase expensive for a large number of indices unless they are almost all similar.

In our examples of `NewUpdate`, we do not implement this optimization.

13.3.5 *Back to Libor*

The VHW implements specific updaters for any indices it knows how to handle; to accomplish this, it must inspect the index. We can use `dynamic_cast`:

```
—————————————— VHW.cpp ——————————————
SDEImp::UpdateOne_* VHW_::NewUpdateOne
    (_ENV, RequestAtTime_& t,
     const Index_& index)
const
{
    if (auto ir = dynamic_cast<const Index::IR_*>(&index))
```

```
      {
         REQUIRE(ir->ccy_ == yc_->ccy_, "Non-domestic rate index");
         const Date_ start = ir->StartDate(t.t_);
10       if (IsLiborTenor(ir->tenor_))
            return NewLiborUpdater(_env, *yc_, *vol_, t.t_,
               ir->StartDate(t.t_), PeriodLength_(ir->tenor_));
         if (IsSwapTenor(ir->tenor_))
            return SDEImp::NewSwapUpdater(t, *ir);
15       THROW("Unrecognized interest rate tenor");
      }
      if (auto df = dynamic_cast<const Index::DF_*>(&index))
      {
         return NewDfUpdater(*yc_, *vol_, t.t_, *df);
20    }
      THROW("VHW model can't simulate non-IR indices");
   }
```

Here `NewLiborUpdater` and `NewSwapUpdater` are local free functions which do whatever precomputation is possible. The definition of
`NewLiborUpdater` given above needs to change only to accomodate the
new definition of `UpdateOne_::operator()`.

In the code above, we pass t rather than just t.t_ to `NewSwapUpdater`.
This allows the swap updater to request that other index values – Libor
rates and discount factors – be available when it is invoked. This decouples
the swap rate computation from the VHW model, so this functionality can
be provided to all SDE's.

For the VHW model, we can use this default implementation as shown
above, or can create our own optimized swap updater (*e.g.*, using cubic-
spline interpolation to get the swap rate directly as a function of S).

13.4   Cox-Ingersoll-Ross

The Cox-Ingersoll-Ross (CIR) model relies on a short rate dynamics of the
form:

$$dr = \kappa(\mu - r)dt + \sigma\sqrt{r}dW.$$

This model has some disadvantages compared to the VHW model we have
just discussed:

- Its parametrization cannot easily be recast into the gH-form, because
  rates remain positive only when $2\kappa\mu > \sigma^2$. Thus a region of negative
  κ, or an increase in H, permits negative rates from which the model
  cannot recover.

- The bond pricing formula is still of the form $P = Ae^{-Br}$, but the computation of the coefficients is substantially more complicated, especially when κ and σ are not constant.
- Implementation of an arbitrage-free Monte Carlo stepper is quite complex (as it is for the Heston model of a single asset price), and the stepper cannot be expected to be highly efficient.
- The $\beta = \frac{1}{2}$ behavior is hard-wired into the model, just as thoroughly as $\beta = 0$ in the VHW model.

On the other hand, it is plausible that $\beta = \frac{1}{2}$ is a better choice, more likely to be close to the market implied elasticity.

For these reasons, we probably would not trouble to implement this model. Should we choose to do so, however, the path is reasonably clear. We will make κ and σ piecewise constant, so that the coefficients A and B of the bond pricing formulas will remain available in closed form. Bond options can then also be priced in closed form, and the PDE coefficients are readily available. The Monte Carlo step requires some degree of approximation.

13.5 Black-Karasinski

The Black-Karasinski (BK) model, which extends the earlier Black-Derman-Toy model with the most flexible one-state-variable dynamics, is usually written

$$d\mathrm{log}r_t = \kappa(t)(\theta(t) - \mathrm{log}r_t)dt + \sigma(t)dW_t$$

where, as before, θ must be time-dependent to fit the yield curve, and κ and σ may be time-dependent to allow a more flexible volatility structure. Note that this is just the VHW dynamics with r_t replaced by $\mathrm{log}r_t$. We can rewrite this model in the form

$$dS_t = g(t)dW_t, \qquad \mathrm{log}r_t = E[\mathrm{log}r] + H(t)S_t.$$

While this captures the features of the "classical" BK model, the short rate elasticity is fixed at one, so there is no gain in flexibility to compensate us for the inevitable loss of analytic tractability. Thus the standard BK model is valuable only if we have some reason to believe $\beta \equiv 1$.

To find a more flexible model form, we start by rewriting the relation between r and S as

$$r_t = \bar{r}(t)e^{H(t)S_t}$$

where $\bar{r}$, like $E[\log r]$, must be fitted numerically to the yield curve. This shows that the lognormal behavior of the BK model is only one choice among many, and can be replaced by a more general functional form. Models based on this idea are often called *generalized Brownian motion* (GBM) models.

What market features should we attempt to model with this new freedom? We can certainly use it to model skew, by giving r the *shifted lognormal* dynamics

$$r_t = \Big(\bar{r}(t) + \Delta(t)\Big)e^{H(t)S_t} - \Delta(t)$$

for a (possibly time-dependent) shift Δ. Adding a quadratic dependence on S will fatten the tails of the distribution and increase the value of out-of-the-money options; but at some point $r(S)$ will no longer be monotonically increasing and the implied yield curve motions will become markedly unrealistic. Similarly, by creating a rich nonparametric dependence of r on S we can hope to fit option prices across a range of strikes, but the resulting models are often plagued by outlandish parameter values and are likely to do us more harm than good.

The pricing code is largely independent of the parametrization, so we do not need to commit firmly to a single choice. We tend to use the term "Black-Karasinski" to refer to any such one-factor model, not only the "classical" lognormal BK model.

13.5.1 *Forward Induction PDE Sweep*

One feature we demand of any parametrization is a parameter – here called $\bar{r}$ – which can be tuned to match the yield curve. We may choose to fit this parameter as part of the pricing process, using forward induction out to the trade's last maturity date, or during the construction of the model, with the last date provided as part of the model specification. We choose the former approach, performing the fit as part of the creation of an SDE_ from the model (see Sec. 10.2.1).

The PDE solver will propagate S_t, and the mapping $r(S)$ will form the ScalarCoeff_ used for discounting. Fitting $\bar{r}$ involves a numerical search, at each time step, to match the observed discount factor $Z(T)$; this fit can be made much more efficient (converging in 2-3 steps) if we are able to compute $dr/d\bar{r}$ at each S.

This leads us to an interface for the abstract mapping:

```
──────────────────── BKImp.h ────────────────────
   namespace BK
   {
      class Mapping_ : public Storable_
      {
5     public:
         double rBar_;
         virtual double R(double S) const = 0;
         virtual double dRdS(double S) const = 0;
      };
10 }
```

Note that each `Mapping_` is valid for only a single time step; but if a parametric structure is shared across time steps, we can store it in a `Handle_` shared across `Mapping_s`.

13.6   Single Equity with Local Vol

Models with deterministic but price-dependent "local" volatility are quite common in equities modeling, partly because they promise to decompose basket options on several equities by computing sensitivity to single-equity options at an appropriate range of strikes. For heavily path-dependent products – most specifically, for those with strongly *path-dependent gamma* – the assumption of deterministic vols is not justified.

When choosing the representation of such a model, we must immediately decide whether to parametrize the local vols or the implied term vols, from which we can rapidly infer local vols using the celebrated Dupire formula. The latter approach, which makes vanilla option prices immediately available, is more common in practice; but we find it distasteful, since it obscures rather than reveals the actual dynamics used in pricing. Here we will parametrize the local vols directly.

For the present, we assume that interest rates are deterministic; we will return to this issue in the next section. Also, we consider only a single equity process, though most structured equity derivatives involve several underlyings.

We first define an abstraction of a local volatility surface:

```
──────────────────── LVSurface.h ────────────────────
   class LVSurface_ : public Storable_
   {
   public:
      virtual DateTime_ VolStartTime() const = 0;
5     virtual double LocalVol(const DateTime_& t, double s)
```

```
          const = 0;
      virtual double IntervalVol(const DateTime_& t_minus,
          const DateTime_& t_plus, double s) const;

10    virtual StepAccumulator_* NewAccumulator() const = 0;
      virtual pair<double, double> UpdateEnvelope
          (StepAccumulator_* accumulator,
           const DateTime_& t,
           double num_sigma)
15    const = 0;
      // ...
   };
```

The `IntervalVol` function is not pure virtual, because we can reasonably implement it with a set of queries to `LocalVol`. The following recipe works reasonably:

(1) Let $\sigma_\pm$ be the local vol at $(t_\pm, s)$.

(2) Let $\bar{t} = \frac{t_+ + t_-}{2}$ and $\delta_t = t_+ - t_-$.

(3) Let $h = \frac{\sigma_+ \sigma_-}{(\sigma_+ + \sigma_-)} \sqrt{\delta_t}$, the central Brownian bridge width.

(4) Let σ_u be the local vol at $(\bar{t}, se^h)$ and σ_d at $(\bar{t}, se^{-h})$.

(5) The annualized step variance is about $V \equiv \frac{(\sigma_-^2 + \sigma_+^2 + 2\sigma_u^2 + 2\sigma_d^2)}{6}$; use $\sqrt{V\delta_t}$ as the step vol.

13.6.1 *Interpolated Vol*

Our favored parametrization will be based on interpolation in a central region, and CEV-like extrapolation at extreme values of S:

```
───────────────── LVInterp.cpp ─────────────────
class LVInterp_ : public LVSurface_
{
    DateTime_ volStart_;
    Interp2::Of_<DateTime_> vals_;
5   Interp1::Of_<DateTime_> loEdge_, loBeta_;
    Interp1::Of_<DateTime_> hiEdge_, hiBeta_;
}
```

The main member, `vals_`, defines the local (lognormal) volatility in some central region. For extreme strikes, we switch to a CEV dynamics; the `loEdge_` and `hiEdge_` are the bounds of the central region, and the corresponding β's define the extrapolation. We allow all these quantities to be time-dependent, though it makes little difference in practice.

This defines the local vol:

```
———————————————— LVSurface.h ————————————————
public:
    DateTime_ VolStartTime() const { return volStart_; }
    double LocalVol(const DateTime_& t, double s) const
    {
        const double lo = loEdge_(t);
        if (s < lo)
            return LocalVol(t, lo) * pow(s / lo, loBeta_(t) - 1.0);
        const double hi = hiEdge_(t);
        if (s > hi)
            return LocalVol(t, hi) * pow(s / hi, hiBeta_(t) - 1.0);
        return vals_(t, s);
    }
};
```

13.6.2 *Derivation from Implied Vol*

Despite our objections, local vols are often hidden behind a global parametrization of implied vols, and derived on demand using the Dupire formula. The derivation of local vols around the time of a discrete dividend payment, required for a production-quality implementation, requires great care. One advantage of this practice is that an override of `IntervalVol` can produce superior results to the purely local default implementation, since it can use the global implied vol information.

13.6.3 *Model and SDE*

Now let us wrap these dynamics in an SDE to support numerical pricing. For simplicity, we will consider only proportional dividend yields, parametrized using a `PWC_` object. Again, the `Model_` and `SDE_` can share an identity:

```
———————————————— LVModel.cpp ————————————————
class LVModel_ : public Model_, public SDE_
{
    double s0_;
    Handle_<YieldCurve_> yc_;
    Handle_<PWC_> divs_;
    Handle_<LVSurface_> vols_;
    // ...
```

The stepper is similar to that of the VHW model (Sec. 13.2), but naturally it must report a space-dependence of the diffusion coefficient. That can be computed using repeated calls to `LVSurface_::IntervalVol`; this is not so bad for PDE's, but to support a production Monte Carlo we will find it necessary to do some precomputation. For instance, the `IntervalVol` could be precomputed at a hundred or so points covering the envelope, and splined to find the appropriate vol for a given simulation path.

The computation of the envelope is also more involved, because of the lack of global information. It is actually best to introduce this as a new virtual member in `LVSurface_`:

```
_____ LVSurface.h _____
    virtual StepAccumulator_* NewAccumulator() const = 0;
    virtual pair<double, double> UpdateEnvelope
        (StepAccumulator_* accumulator,
         const DateTime_& t,
5        double num_sigma)
    const = 0;
```

The call to `NewAccumulator` creates a derived class, visible only to the vol surface, containing information about the envelope so far. Then each call to `UpdateEnvelope` advances the `accumulator` to the new time t, and returns an envelope computed from it. For a Dupire-derived surface no accumulator is needed, and `NewAccumulator` can return 0; for our interpolated local vols, we use a weighted average of vols in the central region and at the existing envelope edge to widen the envelope.

13.7 A Simple Hybrid Model

Now let us combine this local-vol model with our Vasicek-Hull-White model. At the same time, we will consider modeling with more than one equity. For this, we wish to reuse the code of the single-process models we have already created.

Thus we separate the hybrid model into *components* – the VHW and local vol processes – and *correlations*. The latter will be part of the top-level model specification. We also introduce the concept of *polling*, where for a given request – e.g., an index for which a fixing must be generated – we query the components to find one which can honor the request.

The model should store only a single yield curve – otherwise the process of updating the curve, e.g. for a curve delta computation, is gratuitously complicated. Thus we begin by rewriting our local vol model, collecting the description of the equity process into a single component:

```
_____ LVModel.cpp _____
class LVModel_ : public Model_
{
    Handle_<YieldCurve_> yc_;
```

```
      Handle_<LVComponent_> equity_;
5 public:
      LVModel_(const String_& name, yc_, equity_);
      // ...
```

The component, and the asset models it creates, can no longer assume that the state passed to them is entirely theirs. In constructing an asset model, we must supply a *state offset* which will determine where the component's state is stored inside the whole model state.[12]

The parent SDE_ delegates the creation of asset updaters to components, and sends them a RequestAtTime_ which they can use to communicate with other components. Thus, once the step is taken and the new state is available, our existing machinery suffices to support component models. It is best, when creating the updaters, to be sure that the result is unique – that the model does not somehow contain two components which each think they know how to compute a fixing for some given index.

The sharing of information during stepping is harder. For instance, an equity stepper needs to know the discount rate over the step interval, which supplies its drift term. Attempting to reuse the dependency mechanism of NewUpdate would come at a high price in both maintainability – as we are mixing step and post-step code – and efficiency. A more specialized tool is needed.

We observe that there are only a few ways in which the output of one stepper can be input to another: discount rates, FX vols (for quanto drift), and rate vols (for drift of spread rates). An equity price model including the possibility of default would likely be designed as a single component. Stochastic credit and spreads are beyond the scope of this volume; we will describe the mechanism only for discounting.

To ensure that discount factors for the step are available before they are needed, we bring rate-stepping components to the front and execute them first. But foreign rates will require a quanto shift, which depends on knowledge of the FX vol. Thus we need to ensure that the ordering is

(1) The domestic rate stepper.
(2) All FX steppers.
(3) All foreign rate steppers.

---

[12]This offers no protection against components reading or writing out of range. Enforcing such protection through a system of types is possible but expensive at runtime – remember that asset pricing is a code hotspot. In Sec. 13.7.2 we will show a useful protective mechanism.

(4) An additional phase, where the FX steppers are given a discount factor to complete their update.

(5) All other steppers.

After all this, a component sub-stepper requires several inputs not visible to the top-level Monte Carlo:

```
———————————— ComponentStep.h ————————————
class SubStepper_ : noncopyable
{
public:
      virtual ~SubStepper_();
      virtual MonteCarlo::Workspace_* NewWorkspace
          (const MonteCarlo::record_t&)
      const = 0;
      virtual int NumGaussians() const = 0;
      virtual void Step
          (Vector_<>::const_iterator iid,
           Vector_<>::iterator state,
           MonteCarlo::Workspace_* work,
           ComponentDf_* dfs,
           ComponentQVols_* quantos,
           Random_* extras,
           Vector_<Handle_<DefaultEvent_>>* defaults)
      const = 0;
      virtual void ApplyDf(const ComponentDf_& dfs) const {}
};
```

The structures `ComponentDf_` and `ComponentQVols_` contain discount factors and FX lognormal vols (needed for quanto adjustments) over the step period, respectively; they are read by some components and written by others. Steppers for FX components, which will be called before the `dfs` can be fully populated, are expected to override `ApplyDf` and use it to update the FX state once the necessary discounts are known.

The most natural implementation of `ComponentDf_` would be a `map<String_, double>` but this relies on inefficient string comparisons; it is better to reuse the `CreditId_` in Sec. 10.7 to translate the currencies (held by the components) into integer locations (held by the steppers) within a simple `Vector_<>`.

The top-level *master stepper* synchronizes the calls to `SubStepper_s` and communicates a single rolling discount back to the numerical method. It always has a workspace of its own, so that space for the `dfs` and `quantos` need not be reallocated for every path.

13.7.1 *The Case for Components*

Re-implementing a model in this way involves intensive work, and careful setup of generic mechanisms to do simple jobs without too much inefficiency. Given the comparative ease of simply dispatching all the information manually – creating a specialized `LVPlusVHW_` model and hand-writing an optimal stepper for it – it is natural to ask why one should go to the trouble of isolating components.

The answer is that development is a continuous process, and we can make the most progress by opening the door to small successes on one front at a time. Upgrading a single specialized model is a wide-ranging task, and we have nothing to show until it is fully done. But once the connections between components are formed generically, and the role of each individual component is made explicit, then upgrading one kind of component without needing to recode the entire model becomes a faster path to the creation of valuable models yielding a competitive advantage.

A further reason is that a risk measurement system is only as strong as its weakest link; it is fruitless to control interest rate skew exposures in a Bermudan book if we also have a PRDC book where the same exposures cannot even be measured. Component-based models let us refine a process which will give an interesting measure of our portfolio's exposure, and then deploy it to measure *all* the derivative trades with relevant risks.

13.7.2 *State Bounds Checks*

The model queries each of its components for the state size, and builds a composite state joining each component state. By placing internal padding inside this composite state, we can detect most range errors. At the outer (model) level, we form the state including extra elements:

```
──────────────────────── LVHWModel.cpp ────────────
   #ifdef debug
   static const Vector_<> STATE_PADDING(4, DA::NAN);
   #else
   static const Vector_<> STATE_PADDING;
 5 #endif

   Vector_<pair<double, double>> LVHWAccumulator_::Envelope
       (const DateTime_& t,
        double num_sigma)
10 const override
   {
       static const Vector_<pair<double, double>> PADDING
           = Zip(STATE_PADDING, STATE_PADDING);
```

```
15    Vector_<pair<double, double>> retval
         = ir_->Envelope(t, num_sigma);
      for (auto e : eq_)
      {
          Append(&retval, PADDING);
20        Append(&retval, e->Envelope(t, num_sigma));
      }
      return retval;
}
```

Write errors are checked at the model level, by asserting that the padding elements remain equal to DA::NAN (or another extreme constant of our choice). Read errors are checked at the component and asset level, by asserting that a state variable does not have this value.

This protection cannot be used in PDE pricing, since it expands the state with extra variables that have no dynamics. This is not much of a drawback in practice; most hybrid models already have too many state variables for any feasible PDE, and this kind of check is most useful when individual components have multiple state variables – otherwise range errors are not a threat.

Chapter 14

Semianalytic Pricers

All our discussion so far has been centered on generic numerical pricing, where models and trades can communicate at arm's length. We now turn to specialized *semianalytic methods*, which use knowledge of both the trade and model to rapidly obtain the price, or a close approximation thereto.

For example, a lognormal price diffusion with deterministic interest rate and dividend yield leads to the Black-Scholes pricing formula for equity options; and we need solve no PDE in the process. If the deterministic dividend yield is replaced by a set of discrete payments, or a richer parametrization of the dividend curve, then moment-matching methods can yield an inexact but quite accurate price.

As the models grow richer, the semianalytic methods must be correspondingly more sophisticated and complex. This is the "high art" of the quant world. In this volume, we will focus more on the frame: how can we make these approximations available without sacrificing genericity?

14.1 A Moment-Matching Pricer

Consider an equity process, so that the log price is normally distributed at some forward time. If the equity then pays a proportional dividend, the variance of this distribution is unaffected. However, a dividend of fixed size will tend to increase the variance. Suppose that the dividend amount is proportional to S^β for some $\beta \in [0, 1]$; a fixed dividend is simply the $\beta = 0$ limit. If we let $V(t)$ be the accumulated variance of the log at time t and Y the forward dividend yield, we find

$$\frac{\partial V}{\partial t} = \sigma^2 + 2(1 - \beta)VY,$$

plus higher-order terms (*e.g.*, from the higher moments of the true distribution at time t). To a similar approximation, the third moment of $\log S_t$ is governed by

$$\frac{\partial \xi}{\partial t} = \sigma^2 - 2(1 - \beta)^2 V^2 Y.$$

So with these model dynamics, we can use this to price an equity option using only a sum over discrete dividends, or at worst a solution of two ordinary differential equations.[1] We will compute three moments of the distribution of the log; convert them to three moments of the distribution of the stock price at expiry; fit a shifted lognormal distribution to those as in Sec. 7.7.1; and price the option based on this distribution.

```
────────── EquityOptionByMoments.cpp ──────────
     double OptionValue
         (const YieldCurve_& yc,
          double s_0,
          const Dividends_& divs,
   5      const PWC_& vols,
          const DateTime_& expiry,
          const Date_& delivery,
          double strike,
          const OptionType_& type)
  10  {
          const Vector_<> moments = MomentsOfExponent
                 (AccumulateMoments(vols, divs, expiry, 3));
          assert(moments.size() >= 4);    // 0, 1, 2, 3
          double shift, shiftedFwd, shiftedVol;
  15      FitThreeMomentsSLN
                    (moments[1], moments[2], moments[3],
                     &shift, &shiftedFwd, &shiftedVol);
          return Distribution::Black_(shiftedFwd, shiftedVol, 1.0)
                    .OptionPrice(strike - shift, type);
  20  }
```

14.2   Multimethod Objects

How are we to reach this code from a generic function that receives only a `Trade_` and a `Model_`? Because we must check the type of both trade and model (and sometimes their internal properties as well), such a pricing request requires a *multimethod* – a function which is "virtual" in more than one of its inputs. Little optimization is possible, because the internal

---

[1]This is a bad model in several ways - equity skew is not really explained by non-proportional dividends – but it illustrates our approach.

switching process (however we control it) is inevitably irregular; there will be a patchwork of methods with very few sweeping generalities.

Thus we must prepare to query methods in sequence. For this purpose, we will add interface functionality to our trades and models, using mixins; then a given *candidate pricer* can inspect them with `dynamic_cast` to see whether they are suitable to its needs. All candidate pricers derive from:

```
──────────────── Semianalytic.h ────────────────
   struct SemianalyticPricer_
   {
       virtual bool Attempt
          (_ENV, const Trade_& trade,
5          const Model_& model,
           const Valuation::Parameters_& params,
           Vector_<pair<String_, double>>* vals)
       const = 0;
   };
```

The derived classes should contain no member data – for which they can have no valid use – and the lack of a destructor is a reminder of this. For instance, a pricer to attempt the equity valuation above would look like:

```
──────────────── EquityOptionByMoments.cpp ────────────────
   // deprecated, fat interface
   struct PriceByMoments_ : SemianalyticPricer_
   {
       bool Attempt
5         (_ENV, const Trade_& trade,
           const Model_& model,
           const Valuation::Parameters_& params,
           Vector_<pair<String_, double>>* vals)
       const override
10     {
           scoped_ptr<EquityOptionData_> opt
                  (EquityOption::NewData(trade));
           DYN_PTR(myModel, const BlackWithDividends_, &model);
           if (!myModel || !opt.get())
15             return false;
           const String_& eq = opt->eqName_;
           const double value = OptionValue
                  (*model.YieldCurve(trade.valueCcy_),
                   myModel->Spot(eq), myModel->Dividends(eq),
20                 myModel->Vols(eq), opt->expiry_,
                   opt->delivery_, opt->strike_, opt->type_);
           vals->push_back(make_pair
                  (trade.valueNames_[0], value));
           return true;
25     }
   };
```

Here we have showed two ways the pricer can query its inputs: by direct
`dynamic_cast`, or by calling a utility function (`EquityOption::NewData`),
which will presumably cast internally and return the relevant information
upon success.

The resulting code is subpar, because we have made the model disclose
too much of its data, and have used the resulting overly fat interface to
create model-specific functionality located away from the model it depends
on. Continuing along this path would require us to create similar pricers
for each equity model, repeating our mistake.

We can remedy this by creating a mixin for models with semianalytic
pricing of equity options:

```
───────────────── EquityOptionSemianalytic.h ─────────────────
  class HasAnalyticEquity_
  {
  public:
    virtual double Price(const EquityOptionData_) const = 0;
5 };
```

This mixin class will live in the same header file as `EquityOptionData_`;
by deriving from it and implementing `Price`, models will participate in
the pricing service it offers. The candidate pricer should live in the same
component as the mixin class:

```
───────────────── EquityOptionSemianalytic.cpp ─────────────────
  struct PriceEquityOption_ : SemianalyticPricer_
  {
    bool Attempt
      (_ENV, const Trade_& trade,
5      const Model_& model,
       const Valuation::Parameters_&,
       Vector_<pair<String_, double>>* vals)
    const override
    {
10      auto myOpt = dynamic_cast<const IsEquityOption_*>(&trade);
        auto mm = dynamic_cast<const HasAnalyticEquity_*>(&model);
        if (!mm || !myOpt)
            return false;
        vals->push_back(make_pair
15          (trade.valueNames_[0], mm->Price(*myOpt)));
        return true;
    }
  };
```

This code will now be invoked for any model that (by deriving from
`HasAnalyticEquity_`) declares that it can price an equity option. We
prefer the mixin approach, because it allows the specification of a low-level

interface (which will be grouped with the protocols of Ch. 10) which each trade or model can fulfil as it sees fit.

14.3 Method Registry

Now we need a top-level function which can `Attempt` each candidate pricer until one succeeds. We store a run-time singleton registry of such functions, indexed by priority; thus the most common and fastest pricers can be checked first. We use `greater<int>` as the registry's sort criterion, so that large numbers mean high priority.

```
————————— Semianalytic.cpp —————————
Vector_<pair<String_, double>> Semianalytic::Value
    (_ENV, const Trade_& trade,
     const Model_& model,
     const ValuationParameters_* params)
5  {
       const auto& candidates = XThePricers();
       Vector_<pair<String_, double>> retval;
       for (auto pc : candidates)
       {
10         if (pc.second->Attempt(_env, trade, model, params, &retval))
               return retval;
           retval.clear();
       }
       THROW("Can't find any pricer");
15 }
```

The call to `clear` after a failure is needed for composite trades (see Sec. 11.9.2), which may price some components before failing.

Registration requires an instance of the pricer to be tried:

```
————————— EquityOptionSemianalytic.cpp —————————
RUN_AT_LOAD(Semianalytic::Register(new PriceEquityOption_, 5));
```

14.4 Interaction with Re-Evaluator

In Sec. 10.12, we showed a `ReEvaluator_` for stable pricing during bumped runs. Many semianalytic methods need no such stabilization, but any whose execution thread can deviate from a single internal pricing path – e.g., by using an adaptive integrator or ODE solver – will be subject to instability if the bumped run does not exactly follow in the footsteps of its base.

Rather than expand the interface of `Semianalytic::Value`, we pass an auditor through its environment, which will allow candidate pricers to store and later retrieve information.

```
                    ──────── Semianalytic.cpp ────────
    class ReevaluateSemianalytic_ : public ReEvaluator_
    {
        AuditorImp_ auditor_;
        const Trade_& trade_;
 5      const ValuationParameters_* params_;

    public:
        ReevaluateSemianalytic_
            (_ENV, const Trade_& trade,
10           const Model_& model,
             const ValuationParameters_* params)
            :
        trade_(trade),
        params_(params)
15      {
            ENV_ADD(auditor_);
            baseVals_ = Semianalytic::Value(_env, trade, model, params);
            auditor_.mode_ = AuditorImp_::SHOWING;
        }
20
        Vector_<pair<String_, double>> Values
            (_ENV, const Model_* bumped = nullptr)
        const override
        {
25          ENV_ADD(auditor_);
            return bumped
                    ? Semianalytic::Value(_env, trade_, *bumped, params_)
                    : baseVals_;
        }
30  };
```

Now any semianalytic method making a potentially discontinuous change in pricing – such as an integrator deciding how many points to use – can store and later recall its value using the techniques of Sec. 5.5.

14.5   Interaction with Composites

If the components of a composite trade can all be valued semianalytically, then we should be able to similarly value the composite trade itself. We accomplish this with a candidate pricer which calls back into `ValueSemianalytic`:

```
  ───────────────── SemianalyticComposite.cpp ─────────
  struct PriceComposite_ : SemianalyticPricer_
  {
      bool Attempt
          (_ENV, const Trade_& trade,
5          const Model_& model,
           const ValuationParameters_* params,
           Vector_<pair<String_, double>>* vals)
      const override
      {
10        auto composite =
              dynamic_cast<const IsCompositeTrade_*>(&trade);
          if (!composite)
              return false;
          for (auto s : composite->SubTrades())
15            Append(vals, Semianalytic::Value
                  (_env, *s, model, params));
          *vals = composite->FinalValues(*vals);
          return true;
      }
20 };
   RUN_AT_LOAD(Semianalytic::Register(new PriceComposite_, 9));
```

The `CollectionTrade_` and `RemappingTrade_` from Sec. 11.9 will inherit from `IsCompositeTrade_` and override its member functions, `SubTrades` and `FinalValues`.

A sharp-eyed reader may already have noticed the possibility of an unpleasant interaction between `ReevaluateSemianalytic_` and `PriceComposite_`: what if the same pricer code stores auditing information for more than one trade? To avoid this, we have to tag the auditing information with the name or (better) the address of the trade being priced; this in turn requires extra information in structures like `EquityOptionData_`. Alternatively, we can choose a "good-enough" solution where we tag the auditing information with some trade information related to the choice being remembered – *e.g.*, a strike and/or expiry – which will almost certainly be unique in practice.

14.6 Pure Pricers

There are some models – such as equity implied vol surfaces, swaption cubes, and copulas – which do not support any numerical pricing; they simply postulate a distribution of some asset price, in the appropriate measure, at a future time when it will be needed for pricing. These *pure pricers* (also called *effective models*) fit very naturally into our framework: they derive from `Model_` but can never produce an `SDE_`. Thus their entire complexity is marshalled in support of some `Attempt` at direct semianalytic pricing.

14.7 Trade-Dependent Calibration

Our `SDE_` is customized to the trade being priced in a very weak way: it will avoid modeling underlyings which do not contribute to the trade, and may truncate yield curves when possible to save computation. We can envision a system of much stronger trade-dependence, where some detailed features of the trade are used to drive a *trade-dependent calibration* followed by numerical pricing in the calibrated model.

This is a distasteful practice: it reduces the internal consistency of our prices for different products, and distorts our aggregation of market risks across trades. However, if our models are too weak to fit the market with a single calibration, allowing a trade-dependent calibration can mask this weakness.

Models of this sort will produce a new `SDE_` for every trade request, whereas a simple model like VHW will always produce the same `SDE_`. This is why `ForTrade` should return a `Handle_`, which may be to a new or a cached object.

A common example is the USD Bermudan swaption market, which is liquid enough that we can accurately observe prices across a wide range of maturities, first-call tenors, and strikes. We might fit the European market with an effective model simply containing a cube of Black or SABR vol parameters; but we cannot price Bermudans without using an SDE.

A common solution is to calibrate a one-factor model to the *European components* of the Bermudan swaption – the options with the same terms as the Bermudan but restricted to a single expiry – or even to the single most valuable European component only.[2] We would use VHW (Sec. 13.1) for this purpose; thus the model, when asked to create an SDE for a trade, would calibrate a new VHW model and return it as the SDE.

To make this work, we need more information in `Underlying_`. We do not want to add any `Trade_` or `TradeData_` members – that would defeat our purpose of making `Underlying_` a low-level protocol by which a trade can express its needs. But we do need to embed polymorphic information.

---

[2]The latter approach will occasionally show unstable risk, whenever a market change or bump causes a different component to become the most valuable.

Our solution is a narrow interface class, really just a base for dynamic_cast queries, which we will define inside Underlying_:

```
———————————— Underlying.h ————————————
struct Underlying_
{
    class Parent_ : noncopyable
    {
5   public:
        virtual ~Parent_();
    };
    Handle_<Parent_> parent_;
    // continues as before
10 };
```

Most trades will provide an Underlying_ without a parent_, thus not volunteering any information which could drive a recalibration.

A Bermudan option itself cannot tell which component might be the most valuable (it should not even be able to tell which ones have expired), so it must disclose them all regardless of the recalibration being used. We support this with a new kind of Parent_:

```
———————————— BermudanSwaption.h ————————————
class HasEuropeanComponents_ : public Underlying_::Parent_
{
public:
    virtual Vector_<Handle_<Swaption_>> EuropeanComponents()
5   const = 0;
};
```

This will be the base class for a lightweight friend of the Bermudan swaption trade, which can be placed in the Underlying_ it creates.

Once this is written, the pricer can obtain the options for recalibration. Suppose that we have a pure pricer class, SwaptionCube_, and are extending it to support Bermudan options in the manner described. Then we add the code:

```
———————————— SwaptionCube.cpp ————————————
Handle_<SDE_> SwaptionCube_::ForTrade
    (_ENV, const Underlying_& underlying)
const
{
5   auto berm = handle_cast<HasEuropeanComponents_>
        (underlying.parent_);
    REQUIRE(!berm.Empty(), "Can't price non-Bermudan numerically");
    auto euros = berm->EuropeanComponents();
    assert(!euros.empty());
10  auto evaluate = [&](const Handle_<Swaption_>& ec)
    {
```

```
          return Semianalytic::Value(_env, *ec, *this).front().second;
    };

15  auto euroVals = Apply(evaluate, euros);
    const int iMVE = MVEIndex(_env, berm.get(), euroVals);
    const Swaption_& mve = *euros[iMVE];
    scoped_ptr<Model_> vhw
            (VHW::MatchSwaptionHoLee
20                (String_(), YieldCurve(mve.valueCcy_), mve,
                euroVals[iMVE], VolStart())));
    return vhw->ForTrade(_env, underlying);
}
```

This relies on a factory function `MatchSwaptionHoLee` which calibrates a Ho-Lee model's single volatility parameter to one European option.

The `SwaptionCube_` must support some candidate pricer for European options which can be called by `ValueSemianalytic`; the `evaluate` lambda might be changed to call this pricing function directly, rather than meander through `ValueSemianalytic`'s search over candidates. Also, with a little extra code we could check to see whether `iMVE` is found by the auditor, and if so, could avoid pricing the other European components. These optimizations will accelerate Bermudan risk computations by a couple of percent.

14.7.1 *Stabilization*

The helper function `MVEIndex` can simply return the index of the maximum element; but this will lead to occasional explosions in computed risk figures, whenever a small bump happens to cause a different component to become the most valuable. Our stabilization mechanism can prevent this:

```
                        ─── SwaptionCube.cpp ───
int MVEIndex(_ENV, const void* pp, const Vector_<>& euro_vals)
{
    const String_ key = "MVEIndexForBerm" + String::Uniquifier(pp);
    Handle_<IntAsStorable_> store;
5   Environment::Recall(_env, key, &store);
    if (store.Empty())
    {
        const int i = static_cast<int>
            (MaxElement(euro_vals) - euro_vals.begin());
10      store.reset(new IntAsStorable_(i));
        Environment::Audit(_env, key, store);
    }
    return store->i_;
}
```

Here `IntAsStorable_` is a `Storable_ struct` whose sole data member is the index `i_`. The pointer-to-parent `pp` does not affect pricing; it is needed, as described in Sec. 14.5, to ensure that multiple Bermudans embedded in a `CollectionTrade_` will not all try to store their MVE index using the same key.

306   *This page intentionally left blank*

Chapter 15

Risk

To manage a derivatives book, we must measure its exposure to market prices and to less observable parameters – its *risks* – systematically and accurately. If we are to retain flexibility of models, or aggregate risks which are not all measured using the same models, we must ensure that risk is not mapped to model-specific parameters. Thus we emphasize the creation of uniform specifications of a change to a model, which describe a market rather than a model-specific change. Models have the responsibility of responding appropriately to changes thus described; we then think of risk computation as mostly a process of repeated valuation with a series of slightly different models.

15.1 Slides and Bumps

We must specify a change while attempting to avoid specifying the structure of the model to which it will be applied. Such changes are usually described as *slides* – large changes to measure our exposure to a substantial market move – or small *bumps* to measure a first-order sensitivity to a hedge parameter.

In the context of their application, these are indeed different: for instance, the stabilization mechanisms of Secs. 10.12 and 14.4 should always be applied for bumps – so that valuation differences will be only those caused directly by model parameter changes – and never for slides, where the stabilization mechanisms will inevitably suppress the increasingly important higher-order terms. But to a model, this difference is invisible; here we use "slide" to describe a bump of any type or size.

Thus a slide is, in principle, a function which takes one (fully parametrized) input model, and produces a different model. But an implementation such as:

```
// deprecated
class Slide_ : public Storable_
{
public:
5    virtual Handle_<Model_> operator()
        (const Handle_<Model_>& src)
    const = 0;
};
```

will quickly become unworkable, as it forces slides to cope directly with the dynamics of every model we might use. We must instead have the low-level `Slide_` object display its data, and the high-level `Model_` decide how to respond.

For this purpose, it is crucial that *slides must be simple atoms* – in particular, they cannot be decorated or composite objects. If a model determines (probably with `dynamic_cast`) that a given slide affects, *e.g.*, a yield curve, then it should be allowed to assume that the same slide does not also request the manipulation of equity vols. Slides will indeed be composed, but such a composite (which we call a *scenario*) will be represented as a `vector` of individual, atomic slides.

15.2   Mutants

Applying a slide obviously must change the model's contents. We do not want a non-`const` `Apply` function, which would mutate a model in-place; instead, we construct the new `Model_` as a new object. This is the role of `Mutant_Model` from Sec. 10.16.4: The interface (in class `Model_`) is:

─────────────── *Model.h* ───────────────
```
Model_* Mutant_Model
    (const String_& new_name,
     const Vector_<Handle_<Slide_>>& slides)
const;
```

The function is called `Mutant_Model`, not simply `Mutant`, because of the confusion sown by many unrelated methods all named `Mutant`; also, we will often support its implementation in concrete classes with further mutators, which we can then name `Mutant_VHW` etc. It is nonvirtual, relying for its implementation on the pure virtual `private` function

```
────────────────── Model.h ──────────────────
virtual Model_* Mutant_Model
   (const String_* new_name = nullptr,
    const Slide_* slide = nullptr)
   const = 0;
```

The nonvirtual implementation is easy enough:

```
────────────────── VHWModel.cpp ──────────────────
Model_* Model_::Mutant_Model
   (const String_& new_name,
    const Vector_<Handle_<Slide_>>& slides)
   const
5 {
    unique_ptr<Model_> retval(Mutant_Model(&new_name));
    for (auto s : slides)
    {
       unique_ptr<Model_> temp(retval->Mutant_Model(0, s.get()));
10     if (temp.get())
          swap(retval, temp);
    }
    return retval.release();
}
```

The `if` clause allows models to return `nullptr` when mutating for slides that have no effect (*e.g.*, FX slides in a single-currency model). Note that this implementation assumes slides are applied front-to-back, as users will invariably expect.

15.3 Reports

In a given computation, we will extract sensitivities of some set of values (because of composite trades, we must always consider the possibility of multiple values), to some set of parameters. The parameters may be organized along more than one axis – for example, a local vol sensitivity would be indexed by time and by strike. Finally, we may choose to report multiple *views* of the same data as a service to our users; for instance, we might report both a raw sensitivity and the notional amount of an instrument required to hedge it. Thus a risk report must organize information along multiple axes – up to at least five.

Though the report's dimension is unknown, we store its entries in a single container; we use `deque` rather than `vector` to avoid demanding a vast amount of contiguous memory.[1] Thus we begin our implementation with

---

[1] The machine-generated `Write` function relies on `Vector_` entries. Thus during serialization we might end up with three copies of the data (the `deque`, the vector, and the serial

```
──────────────────── Report.h ────────────────────
class Report_ : public Storable_
{
   map<String_, int> axes_;    // lookup location
   Vector_<int> strides_;      // back() is whole size
5  deque<double> vals_;

public:
   Report_(const String_& nm, const Vector_<Report::Axis_>& axes);

10    void Write(Archive::Store_& dst) const override;
   Vector_<String_> Axes() const { return Keys(axes_); }
   int Size(const String_& axis) const;

   typedef Report::Address_ Address_;
15    Address_ MakeAddress() const;
   double& operator[](const Address_& loc);
   const double& operator[](const Address_& loc) const;
   // ...
```

The layout is elucidated by a look at `Size`:

```
──────────────────── Report.cpp ────────────────────
int Report_::Size(const String_& axis) const
{
   REQUIRE(axes_.count(axis), "No axis '" + axis + "'");
   const int which = axes_.find(axis)->second;
5  return (which > 0 ? strides_[which - 1] : vals_.size())
      / strides_[which];
}
```

The `Report::Axis_` includes a name, which we translate to an index using the lookup `axes_`, and a size. Axes with larger indices walk through `vals_` with successively smaller `strides_`. We will define `Axis_` fully after discussing headers and bookkeeping information, in Sec. 15.4.

We encapsulate element access in a single-argument `operator[]` which accesses a single `Address_` value; there are two viable implementations. The easiest way is to define an address which is essentially a `map<String_, int>`; this supports idioms like:

```
for (loc[TRADE] = 0; loc[TRADE] < nTrades; ++loc[TRADE])
```

where `loc` is the address. At each call to the report's `operator[]` (not the address's), we will translate the string keys to corresponding axes.

It is slightly more efficient to have the address type store the axis ordering; then the lookup occurs only in the address's `operator[]`, which is called less often. The catch is that we cannot construct such an address until we have an instance of the report — this is the role of `MakeAddress`.

```
────────────── Report.h ──────────────
namespace Report
{
   struct Address_
   {
      map<String_, int> axes_;
      Vector_<int> locs_;
      Address_(axes_) : locs_(axes.size()) {}
      int& operator[](const String_& axis);
   };
}
```

The translation of the `Address_` into a location inside the report is quite efficient:

```
────────────── Report.cpp ──────────────
const double& Report_::operator[](const Report::Address_& I) const
{
   const int which = InnerProduct(I.locs_, strides_);
   REQUIRE(which < vals_.size(), "Invalid index into report");
   return vals_[which];
}
```

This storage scheme allows us to resize the report along the first axis, by resizing `vals_` appropriately, without moving any existing elements; we implement this refinement in the non-`const` version of `operator[]`. This is convenient when trades can have multiple values; rather than count the number of values before creating the `Report_`, we simply ensure that the trades are along the first axis. For axes beyond the first, we must still supply the correct size to `Report_`'s constructor.

15.3.1 *Barewords*

We have used `TRADE` as a variable name above; this is actually the name of a constant defined by:

```
────────────── ReportUtils.h ──────────────
namespace ReportAxes
{
   BAREWORD(TRADE);
   BAREWORD(VIEW);
   // ...
}
using namespace ReportAxes;
```

This makes our meaning clear while enlisting the compiler to check for misspellings.

15.4   Portfolios

So far we have talked about individual trades; also, we have discussed only the financial content of trades, ignoring bookkeeping issues such as the need to record the trader and counterparty. Also, we have talked about valuation parameters without describing their origin.

We will use *portfolio* to mean a collection of trades, plus additionally for each trade:

- Bookkeeping information, which we may need to reflect in a risk report.
- Valuation parameters controlling the running of risk.

A portfolio, as stored by a system, will likely associate further information with each trade, such as a rule for finding the model with which to value it.

Since bookkeeping information must be communicated from a `Portfolio_` to a `Report_`, we add a place to hold it:

```
                          ──────── Report.h ────────
   namespace Report

   {
      struct Header_
5     {
         Vector_<String_> labels_;
         Matrix_<String_> values_; // one column per label
      };
   }
10
   // ...

   class Report_
   {
15 // ...
      Vector_<Report::Header_> headers_;
   public:
      void AddHeaderRow
         (const String_& axes,
20       int offset,
         const Vector_<String_>& values);
      // ...
```

which the `Portfolio_` can write to. Thus we need:

```
————————————— Portfolio.h —————————————
class Portfolio_ : public Storable_
{
    Vector_<Handle_<Trade_>> trades_;
    Vector_<ValuationParameters_> params_;
5   Vector_<TradeBookkeeping_> books_;
public:
    int NTrades() const { return trades_.size(); }
    Handle_<Trade_> Trade(int i_trade) const;
    void Clear(int i_trade) const;
10  const ValuationParameters_& ValuationParams
        (int i_trade) const;

    Vector_<String_> Bookkeeping(int i_trade) const;
    Vector_<String_> BookkeepingLabels() const;
15 };
```

The labels will be provided when constructing the report. Thus we finally know the specification of the `Axis_`:

```
————————————— report.h —————————————
namespace Report
{
    struct Axis_
5   {
        String_ name_;
        int size_;
        Vector_<String_> labels_;    // will be placed in a Header_
    };
}
```

15.5   Tasks

Having described what must be the output of a risk run, we examine its inputs. It must take a portfolio and model; in the process of pricing, these may make some demands on the environment (*e.g.*, for the presence of certain fixings).

Thus we define a generic risk task:

```
————————————— Risk.h —————————————
class RiskTask_ : public Storable_
{
public:
    virtual Report_* Run
5       (_ENV, const Portfolio_& portfolio,
        const Model_& model)
        const = 0;
};
```

This class must be `Storable_` because a call to `Run` is the atom of a distributed computation.[2]

15.6   Slide Utilities

Many objects will be shared across model types; for slides which can be implemented as changes to these objects, we place this implementation in a utility function for models to call as needed.[3] The canonical example is the yield curve, shared by essentially all models. For this we provide a shared utility:

```
——————————————————— SlideIR.h ———————————————————
namespace SlideIr
{
    SlideEffect_ Apply
        (Vector_<Handle_<YieldCurve_>>* curves,
5        const Slide_& slide);
}
```

The subtlety here is that there are two ways in which the slide can have no effect on curves: if it is not an interest-rate slide, or if it slides some curve not present in the model. In the former case, the model must continue inspecting the slide to see if it affects some other parameter. We create an enumerated type (for once, without mark-up) to allow utility functions to return this information:

```
——————————————————— SlideUtils.h ———————————————————
enum class SlideEffect_ : char
{
    UNKNOWN,
    ABSENT,
5    NO_OP,
    EFFECTIVE
};
```

Thus the `UNKNOWN` case is used to signal that the model must continue checking slide types. The calling code will look like:

```
if ((eff = SlideIr::Apply(&newCurves, slide)) !=
    SlideEffect_::UNKNOWN)
{
    changed = changed || eff == SlideEffect_::EFFECTIVE;
5    // act on changed curves
}
else
    // check some other kind of slide
```

---

[2] We can sometimes split the `Run` itself, but generic techniques for this are beyond this volume's scope.

[3] And *never* in base classes for derived classes to call.

Since slides will be accumulated to effect a scenario, this code will likely be inside a loop over slides, and if `Apply` succeeds we will simply `continue`.

15.7 Conclusions

Slides provide us with a general way to perturb any model that can respond to them; thus we can construct risk computations without direct reference to the model type. The techniques of Ch. 10 let our trades communicate with *any* model in numerical pricing, allowing us to prototype new models on existing trades and perform alternative risk analyses. Ch. 14 shows how we can insert fast semianalytic methods wherever they are appropriate, and preserve the stability of risk.

We have worked carefully to separate trades and models; to enable component-based models for prototyping and hybrids; and to provide flexible methods for preserving stability of risk regardless of the details of pricing. These are crucial steps on the road to analytics superiority.

Chapter 16

Appendix: The Age of Stochastic Calculus

The following is adapted from my prepared text for a speech to a plenary session of the ICBI Global Derivatives conference in 2012, an audience of quant practitioners and academics.

The Age of Stochastic Calculus is ending. We have been guided through our entire professional lives by a unifying idiom, of pricing and hedging based on mathematical analysis of compactly represented models – but that idiom is less fundamental, and less necessary, than we have thought.

The central insight driving the derivatives markets is that we can price a wide range of payouts, accurately enough to make markets in them, by relating them to the payout of a set of hedge instruments or a hedging strategy – for which in turn we can find prices in a "less derived" market. In practice, this usually involves dynamic as well as static hedging: thus we must somehow estimate the future value of trades we cannot yet execute.

Our time-honored tradition is to perform this estimation using an arbitrage-free mathematical model of the relevant underlyings, then apply the identity between the cost of replication and the expected payoff. What I want to emphasize is the way this tradition combines two steps of the valuation process. First, it identifies the value of a derivative as the cost of remaining tolerably hedged in a realistic range of scenarios; second, it dictates that those scenarios should be derived from a compact simplified representation of the market – what most of us would call a *model*. These are the two axioms that define the boundaries of our professional world.

The first axiom is the inevitable *sine qua non* of derivatives markets. But the second is not; it is a cultural convention, created in a quest for the simplicity of physical law, reinforced by some high-profile successes, and perpetuated by our own fealty to its worldview – to *our* worldview.

Another near-universal assumption of modeling – a cultural axiom, if you will – is that our models will evolve from a well-defined initial state, a single point in parameter space. Hidden state is anathema to our worldview, and mixture models are repulsive in an almost instinctual way. (This is a valid instinct, by the way; we'll return to this issue later.) But the idea that pricing must start from a precise snapshot of everything is clearly an excessive demand; can we really need that much?

At this point I should explicitly state that I am not challenging the validity of mathematical modeling, or attempting to insinuate that we have collectively wasted the last 35 years. Pricing by way of model building is certainly legitimate; however, it is neither unique nor inevitable.

The idea of mathematical modeling, besides its practical merits, is aesthetically attractive to us; to the graduate students we recruit; and to clients, who want evidence that investment banks are providing them with the fruits of real expertise. Mathematics will always beat technology in the swimsuit competition.

And the strongest, most influential quants have reinforced our mindset. Recognizing that oversimplification is the major shortcoming of our models, they have developed increasingly sophisticated and sexy methods to increase their realism and flexibility. Their visible success guides our work, and is in turn shaped by how we, as a group, define success.

As the scale and complexity of derivatives trading have increased, the computing power dedicated to these models has grown commensurately. This provides a crucial operational advantage to mathematical models: they provide a compact description of the world, an easily communicated kernel, which makes the bandwidth of distributed computing affordable. Our modeling paradigm accepts increased local CPU cycles in exchange for decreased bandwidth; this was a good trade until at least about 2005, and now it continues because it has shaped our business's perception of what computers are for.

(I personally tend to think of valuation in terms of Monte Carlo, but this is not really relevant here; some more efficient method might be available, but in any case our pricing will tend to intensively use local CPU.)

Let me spend a couple of minutes reviewing the changes – in the world we are trying to model, and in the available tools and technologies – that are undermining our old traditions. First there is the loss of fungibility; swaps with different counterparties can no longer be considered as equivalent instruments, so opposite swaps no longer have fully offsetting risk. I suppose Vladimir Piterbarg and his colleagues at BarCap are leading the way here, both in formalizing the leading-order framework and in exploring higher-order effects.

Central clearing does provide fungibility, but we should understand that it does not really remove counterparty risk – most realistic scenarios for the failure of a major trading bank would bring down the exchange as well, or allow losses to pass through to counterparties. An exchange, if it trades any products which can put a major bank at risk, becomes a weak credit, a canary in the coal mines of finance.

A more severe problem is loss of separability: *i.e.*, the breakdown of the simplifying assumption that a trade's value can meaningfully be computed in isolation from the rest of our portfolio. (For decades quants have investigated market impact models, which introduce quadratic functions of notional into the price of derivatives; though somehow the logical consequence never interfered with the quest for scalably popular derivative structures.) More recently, the increasing focus on CVA and RWA has forced us to analyze option-like payouts on baskets of future values. Suddenly we cannot price any single derivative without knowing not only the values, but the entire distribution of future values, for other trades – which may involve completely different underlyings.

Different asset classes have dealt with these dilemmas in their own ways. In the traditional quantitative strongholds of Rates and FX, it is treated as one more thing to be modelled – since most trading banks have the wherewithal to create a simple multi-currency model, and few can move a more sophisticated cross-currency model into full production, this ends up reinforcing the already-dominant methods. In Equities, the standard is apparently to ignore the issue, which is possibly justifiable; product durations are shorter and, while the rates derivatives market is to some extent a source of lending, the equity market generally is not. In Credit, our need for dynamics to simulate the distribution of future values wars with our need for copulas to price large credit baskets, and no equilibrium has been achieved.

The salient point here is that we are not approaching any ideal model: the ideal is receding from us further into the future. Our toolkit is growing and maturing, but not keeping up with the problem we face.

Let's change course now, and think about hardware for a minute. Your business probably uses a distributed computing grid to marshal enough CPU power to run risk on its portfolio. Total available CPU is itself probably the limiting factor in what you can compute, unless you have a bottleneck at a badly written central distributor. We need CPU, rather than bandwidth or memory, *because of* the way we represent models (as descriptions of stochastic processes with relatively few parameters, at most a few thousand); this representation ensures that the model will be small, compared to the scale of current data transmission capabilities, without our having to think about it. So in this paradigm, where the thing transmitted to the remote processor is an innately compact model, our assumption that CPU is the sole limiting factor is justified.

An alternative approach, attempting to minimize CPU cost, would be to simulate the paths only once, then transmit the pathset to the remote processor. As an optimization, this fails in most cases. There are also a couple of technical difficulties, around choosing the event times of the simulation and maintaining consistency when pricing products with different but not disjoint underlyings; these are manageable with a little ingenuity, but there has not been much interest in this problem because the payoff seems so small.

With this in mind, let's go back to the fundamental justification for the price of any derivative: the price is driven by the expected cost of reproducing the necessary payoff, in a broad range of scenarios consistent with the market. These scenarios could be Monte Carlo paths from a model, or nodes in a PDE grid with its implied transition probabilities. Or they could be something much broader, and much more reflective of the breadth and variety of possible outcomes. When you stop to think about it, the idea that you should value a security by averaging over thousands of paths, but that these paths should all emerge from a two-currency Hull-White model, is simply risible. We need to keep the idea of pricing by computing the expected payoff over a pathset, but escape from the idea that this pathset emerges from a compact toy model of the world.

Valuing a portfolio this way is extremely bandwidth-intensive. A Brownian motion, or any similar process we might substitute for it, actually contains infinitely many bits of information (which, it could be argued, will exceed the capabilities of our hardware). In practice, we will keep the concept of a Brownian, or Levy, bridge from which we can pull short-timescale information on demand. By creating these bridges deterministically (for instance, using the name of the driver to generate the random seed) we can achieve consistency across processors. We'll generate and share the paths at monthly or quarterly pillar dates, and bridge between those locally.

Let's first think about the task after the pathset has been generated, when we set forth to price trades with it. For concreteness we'll consider 10^5 paths, which is maybe slightly low but probably acceptable; 1000 underlyings, which is high for fixed income but reasonable for equity; and 100 pillar dates. This gives 10^{10} outcomes: too much to store in memory on one machine, but very easy to store using a clustered cache. In fact, the cache can be small enough that the processors involved share a single backplane, and can exchange data peer-to-peer at gigabytes per second. Notice that we are (as of 2012) near the edge of feasibility; eight or ten years ago, we could not have mustered the bandwidth to attempt a project like this.

The valuation process for a single contingent claim now has three main steps. First we fetch the necessary information at the relevant pillar dates; then we bridge to obtain the realizations needed to compute the pay-out; finally we generate the resulting cashflows. Each of these steps takes seconds, rather than minutes, though the last step can be longer if we have a slow scripting language for nonstandard payouts.

At the end of this process, we have the payment stream from each contingent claim on each path. The most obvious thing to do is to collapse this down to a value for each path, then average these to value the derivative. That's a valid output, but we have collected far more information than that.

A wide range of scenarios can be implemented as reweightings of the pathset. In particular, a perturbation to any market price for a vanilla hedge instrument, with other instruments fixed, can be found through a reweighting – for instance, that with minimal change to the weights in the L2 norm given the desired price perturbation. (Maybe you don't like the L2 norm, because it could give negative weights. An old paper by Marco Avellaneda is useful here: he showed that if the weights are uniform and the perturbation is small, then the computed hedge ratio is insensitive to the details of the reweighting algorithm.) The result is that, in the process of valuation and without any extra work, we have also computed the hedge (to first order) into any instrument we care to specify.

(Some of you are preparing to object to the use of control variate weighting, because it introduces arbitrage into the pathset. I think this is going to be a manageable problem – more precisely, I think it is going to be far less severe than the problems we currently face, of placing zero value on outcomes we know are possible but which fall outside the scope of our model. For example, we could monitor the prices of a chosen set of trading strategies to detect any significant nonzero return, which would signal that arbitrage had crept into our model. As long as the "generating model" which produces the paths is close to the "market model" which determines the target prices – *i.e.*, as long as our calibration is reasonably accurate – the arbitrages induced should remain negligible. Mixture models whose separate parts are not arbitrage-free will fail this test, because we can write strategies which estimate their generating state from market observations, then trade based on that knowledge. If the parts are arbitrage-free, we have to be careful that forward implied volatilities do not collapse as the state becomes known.)

The task of generating the pathset is considered independently of pricing. While not trivial, it does not really introduce any new challenges. We can choose a single model or a mixture of models to generate the pathset, and the process is inherently parallelizable. To the extent that we use a very complex generating model, in which we have no hope of closed-form pricing, we have a procedural problem in updating its parameters to keep its pricing reasonably close to the market. We can use control variate to go the last mile, as long as the starting point is good enough that this won't introduce significant arbitrage.

Returning to the output from the valuation using the pathset; suppose we don't collapse in the time dimension, but instead retain the payment profile of the derivative, or collect a net payment profile for a portfolio of derivatives. This is what we need for a CVA or RWA computation. Thus we can see the promise of a world where CVA, instead of being evaluated with simplistic models, is evaluated with models superior to those we are now using for front-office pricing.

We can also begin to quantify the concept of unhedgeable risk. A derivative plus all its first-order hedges – the best hedged portfolio we can create today – will still have some dispersion of values across the pathset. (By the way, static hedging at a barrier, for which I have no respect, can nonetheless be incorporated in this framework: it is just another payout which can in theory be sourced in the market and given a market price.) The residual dispersion of values, even with the hedge, is risk that must be dynamically hedged in the future, thus introducing a dependence on yet-to-be-realized future behavior.

This is just a sketch, which does not do justice to the complexity of a fully working system; but it shows that, with today's technology, we can replace the CPU-intensive pricing based on simplistic models with a new approach that will do a better job of protecting our employers in a wider range of market scenarios. There may be other, better ways to reach the same goal; this is just the one I know.

With all that said – with the end of the Age of Stochastic Calculus in sight – let me spend a few minutes on a more personal level. I started by pointing out the cultural axiom of quants: that we price and hedge based on compactly represented models. This is historically understandable, since this trail has been followed throughout the 40-year rise of the derivatives markets. But there is an emotional reason as well. We would like to agree with Keats that "Beauty is truth, truth beauty;" or, stated with more precision, that "what the imagination seizes as Beauty must be truth." And by beauty, we mean the character of physical law.

But in our world of prediction algorithms, screen painting and dark pools, physical law is unattainable. We are not a Second Foundation, analyzing markets scientifically from afar, but peers scrambling for a shred of competitive advantage. We need to borrow a different phrase from Keats, and cultivate "Negative Capability, that is, when a man is capable of being in uncertainties, mysteries, doubts, without any irritable reaching after fact and reason." We cannot impose law and order on the market processes that threaten us. Our business is not the austere pursuit of truth, but the grimy minimization of error.

I may already be considered something of a Philistine. But I am not blind to the appeal of our Quixotic search. I spent seventeen years on this voyage, and it was often very beautiful. I will mourn this age as much as anyone, when it passes; but I will prepare for the next one.

This page intentionally left blank

Acknowledgements and Further Reading

The standard text on physical code structure is John Lakos's *Large-Scale C++ Software Design*. The principles it sets down are echoed throughout this work.

The uses and ramifications of automatic interface generation were largely worked out by Ian Taylor at Bankers Trust in 1996-97.

Advanced techniques for numerical and semianalytic pricing are derived in Alex Lipton's *Mathematical Methods for Foreign Exchange*.

The routines of *Numerical Recipes*, by Press *et al.*, are a good place to start for many implementation problems. This is also a book to be explored, to learn what methods might be available for problems one has not yet met.

The challenge and promise of higher-level programming are often explored by advocates of non-C languages. I have particularly profited from Paul Graham's *On Lisp*; the *Objective Caml Manual* of Xavier Leroy *et al.*; Mark-Jason Dominus's *Higher-Order Perl* and the Perl 6 language specification; and Andrei Alexandrescu's *Modern C++ Design*.

I learned the importance of the separation of asset and payout from Neil Smout and Trevor Chilton at UBS, who laid the conceptual foundations of Ch. 10.

I have discussed underdetermined root search for calibration of term structures more fully at the ICBI Global Derivatives US conference in 2014.

On a more personal level, I would like to thank many colleagues and friends, who have contributed in more diffuse but no less real ways; especially Leif Andersen, Jesper Andreasen, Christophe Chazot, Steve Dugdale, Andy Felce, Richard Gladwin, Don Goldman, Fabien Hure, Stewart Inglis, Sandeep Jain, James Kanze, Alan LeGuen, Alex Lipton, Chris Mitchell, Vladimir Piterbarg, Dmitry Pugachevsky, Gerson Riddy, Paul Romanelli, Gleb Sandmann, and Richard Waddington.

326 *This page intentionally left blank*

Index

332 *This page intentionally left blank*

Printed in the United States
by Bookmasters

Printed in the United States
By Bookmasters